I0795849

IMAGES
of America

The Aurora Farmers Fair

100 Years

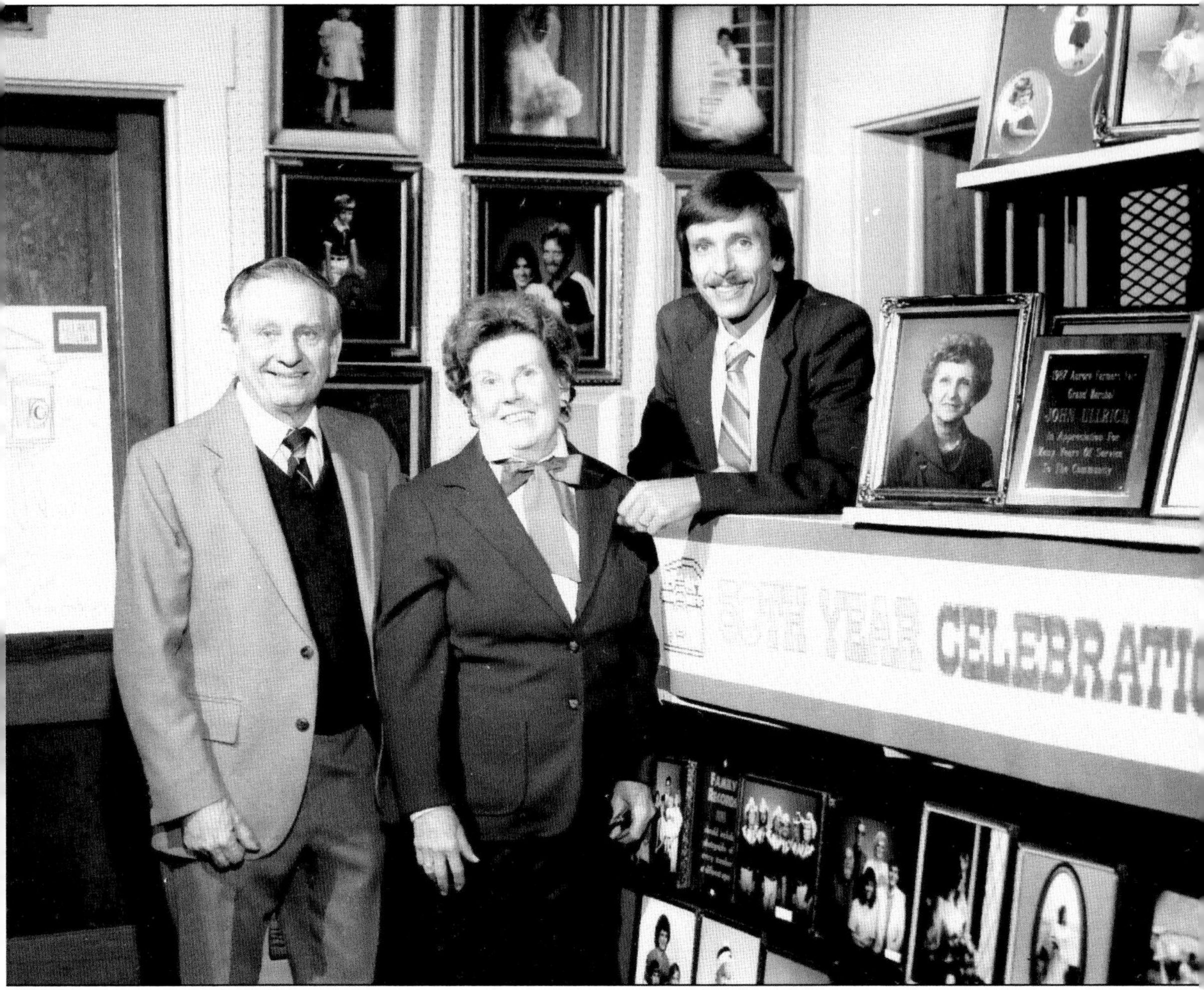

Seen here, from left to right, are John F. Ullrich, sister Rita Ullrich, and son Roger Ullrich. For more information on this family, please see page 6.

On the cover: Representing the Aurora Tri Kappa, Judy (Althoff) Laker glides along the 1954 parade route as a southern belle. Under the dress was a truck driven by Fred Hueseman. Hueseman was able to drive by looking through the shear material on the front of the dress. As if going through a maze, Althoff crawled under the skirts of the dress then onto the truck bed where she then came up through the top of the skirt and sat on the roof of the truck. (J. F. Ullrich collection.)

IMAGES
of America

THE AURORA FARMERS FAIR

100 YEARS

Jenny Awad for the Aurora Lions Club

ARCADIA
PUBLISHING

ISBN 978-0-7385-5168-5

Published by Arcadia Publishing
Charleston SC, Chicago IL, Portsmouth NH, San Francisco CA

Printed in the United States of America

Library of Congress Catalog Card Number: 2007931943

For all general information contact Arcadia Publishing at:
Telephone 843-853-2070
Fax 843-853-0044
E-mail sales@arcadiapublishing.com
For customer service and orders:
Toll-Free 1-888-313-2665

Visit us on the Internet at www.arcadiapublishing.com

This book is dedicated to all who have contributed their time and talents to the ongoing work and success of the Aurora Farmers Fair.

Contents

ACKNOWLEDGMENTS

This book could not have been published if it were not for the photographs of John F. Ullrich and family. Educated only by a home-study course in photography, John started taking pictures when he was in high school using his dad's box camera. In 1935, he took the first pictures of the Aurora Farmers Fair parade out of the Ullrich Drug Store window. John and his sister, Rita Ullrich, developed photographs in a darkroom set up in a closet at the home of their parents. In 1937, they purchased the Whitney-Rullman Studio, on the corner of Third and Bridgeway Streets. In 1946, the brother and sister team relocated to the second floor of Ullrich's Drug Store, on Second and Main Streets. John photographed and documented well over a half century of Aurora and Dearborn County history. Son Roger joined the studio in 1981. John F. Ullrich peacefully passed away March 1, 2007. The Ullrich family generously donated hundreds of Aurora Farmers Fair photographs to the Aurora Lions Club. Other photograph donations came from Barbara Craft, Agnes Teaney Gresham, Jennifer Metcalfe, Orville Green, Ken Strasemeier, Ed Kerr, Dale and Marilyn Moeller, Violet Ravenscraft, Sue Karsteter, the Dr. Leslie Baker family, Jack Tandy, Marge Waldon, Dave Hizer, Carol Chapman, Roberta Anderson, Derald and Phee Ellinghausen, Alma Trennepohl, Tom and Peggy Largent, Bob and Peg Sanford, Suzanne and Dick Ullrich, Charles Johnston, Millard and Sally Rullman, Richard Ullrich, Roberta Anderson, South Dearborn High School, Joe Awad, Register Publications, Aurora Journal Press, Nelson Elliott, Howard Aylor, Doris Smith, and Visions Photo.

INTRODUCTION

This is a book about Aurora's love of a fair—the Aurora Farmers Fair. The story began in 1908 when a newcomer to Aurora, Clarence B. Wilson, began talking to other businessmen about an event to salute the local farmers who were such a vital part of Aurora's economy. Those who were responsible for laying the groundwork for the first fair in 1909 were Edward Chambers, business owner and chicken fancier; William Ketchum, prominent Dearborn County farmer; Joseph R. Houston, Aurora Public Schools superintendent; Adam Hill, wharf boat owner and coal merchant; and William Hoskins, Aurora manufacturer. It is doubtful that their thoughts at the time went beyond that Saturday, October 9, 1909, but as this book shows, what they created is a century of memories—walking the midway and riding the rides or playing one of the many games, watching or marching in either the pet parade or Saturday's giant street parade, or sometimes just gathering with friends and family or renewing old acquaintances.

The book is a collection of photographs that span from that first fair in 1909 to the 99th fair held in 2007. Some are from the archives of the Aurora Lions Club, sponsor of the fair since 1959. However, most are from past and current residents of Aurora documenting the love this great annual event holds in the hearts and minds of those who gather each October to create and partake in the fun, food, and festivities that are the Aurora Farmers Fair.

The story told is the result of combining historical text and informational data with oral histories that trace the fair from that 1909 October Saturday to the present. The materials are in 20-year periods: 1909–1929; 1930–1949; 1950–1969; and so on. Each picture includes the available information. The collection of pictures and related facts used are the responsibility of the committee, which worked so hard to make this book a reality. We regret any errors in historic data, authorship, or names and dates. We hope that any such mistakes will not affect the use and enjoyment of this book.

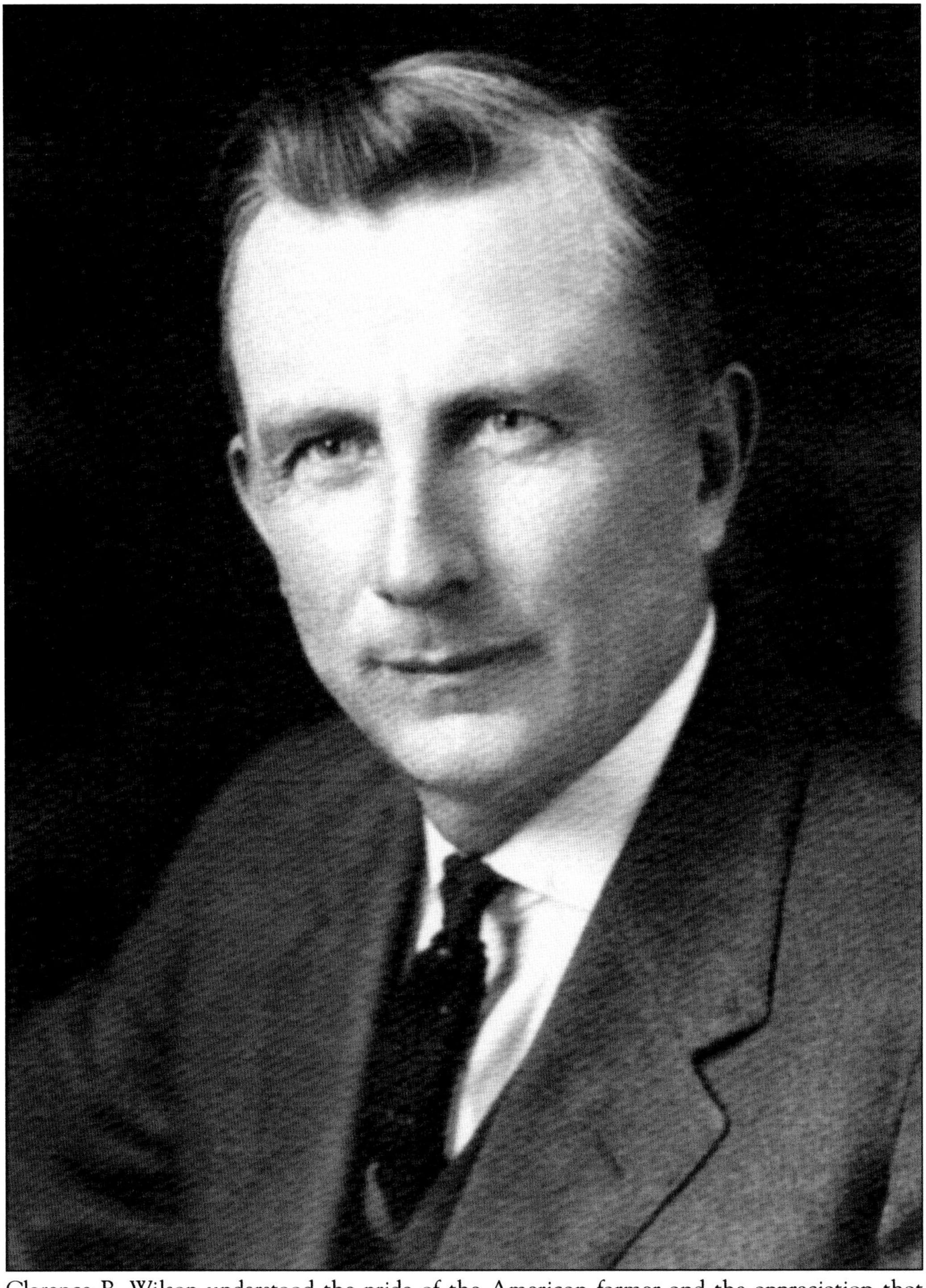

Clarence B. Wilson understood the pride of the American farmer and the appreciation that the Aurora Businessmen's Association would show them by sponsoring many competitions. As president of the Aurora State Bank, he led a group of visionaries into creating an event that will continue as long as there is an Aurora. (Ellinghausen collection.)

One

FROM 1909 TO 1929

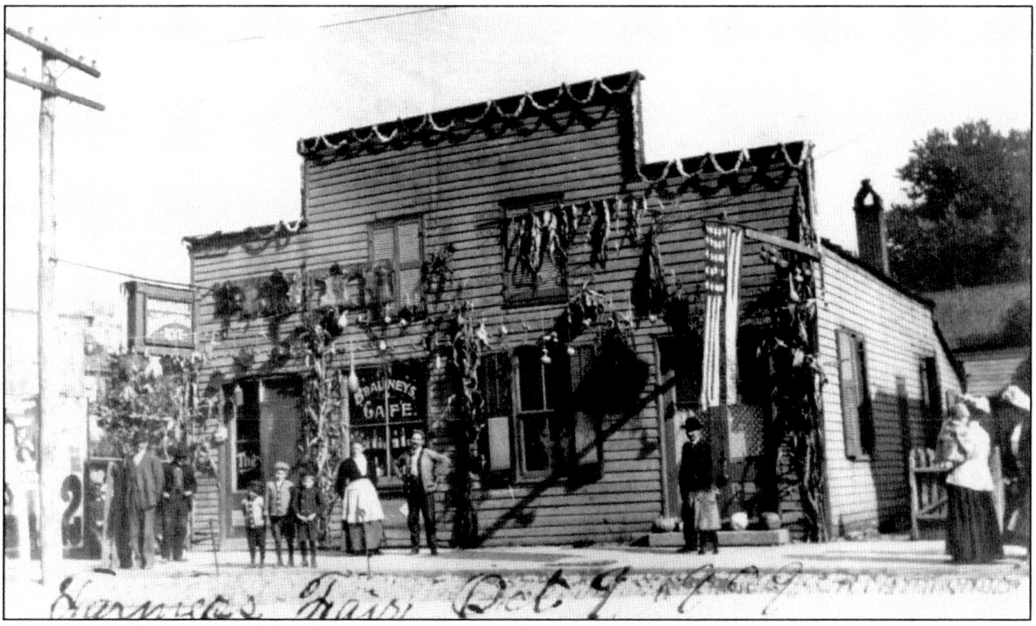

Keeping in the spirit of the fair is Brauney's Café owned by Mike Braunagel located on the site of the present Beyers Pharmacy on Third Street. The café front was graced with hides of many animals, gourds, and corn husks while the interior was covered with the bark from trees. The lady with the apron and the gentleman next to her are the Braunagels. The photograph was taken on October 9, 1909. (Beyer's Pharmacy collection.)

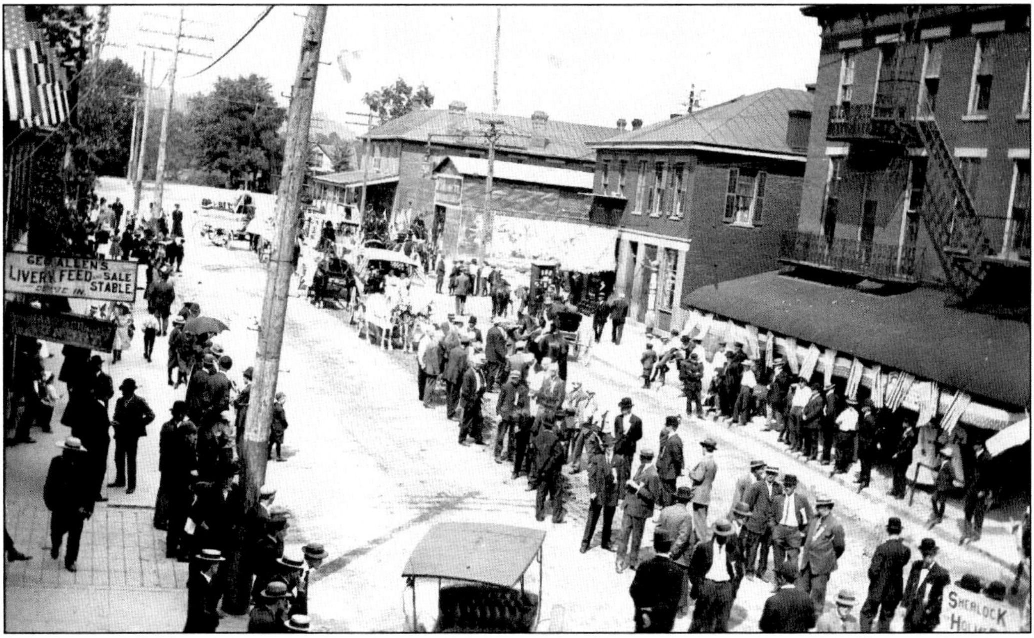

Men of the Mutual Aid and other organizations wait on Second Street for the parade to begin. The parade formed on upper Second Street near the Kirsch House Hotel. The J. N. Kinnett Livery barn was next to the hotel. The Campbell House Hotel is shown at the extreme right. The dark sign at lower left designates the Chinese laundry. George Allen's Livery and Feed was down the alley. (Hizer collection.)

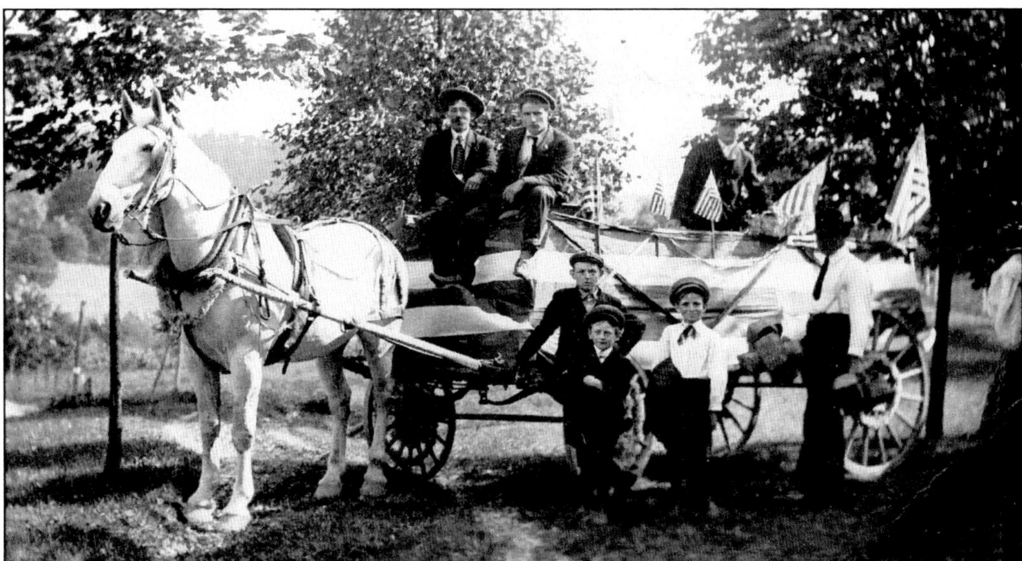

Frank Green and Andy Johnson sit on top of a buckboard owned by the Wooden Shoe Furniture Company. In the photograph above this wagon is rounding the corner at the end of the parade route. (Green collection.)

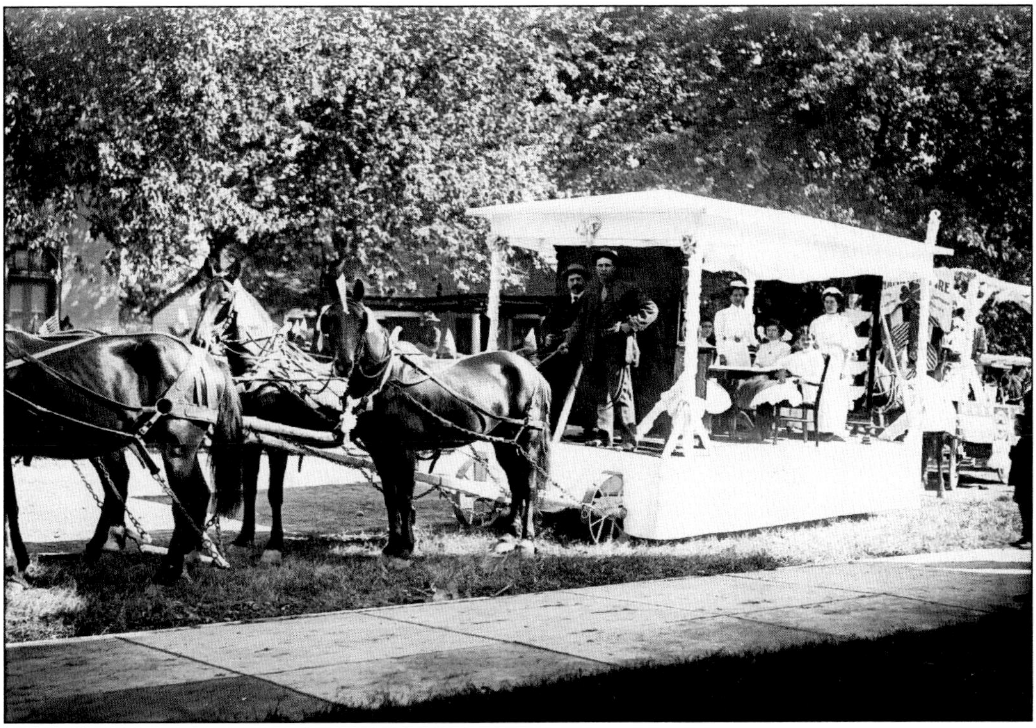

A family in a kitchen with new furniture is displayed on this Aurora Furniture Factory float. Standing, Frank Green and Andy Johnson drove the four-hitch team. The back two horses, Flora and Doc, were on loan from Jim Trester. (Green collection.)

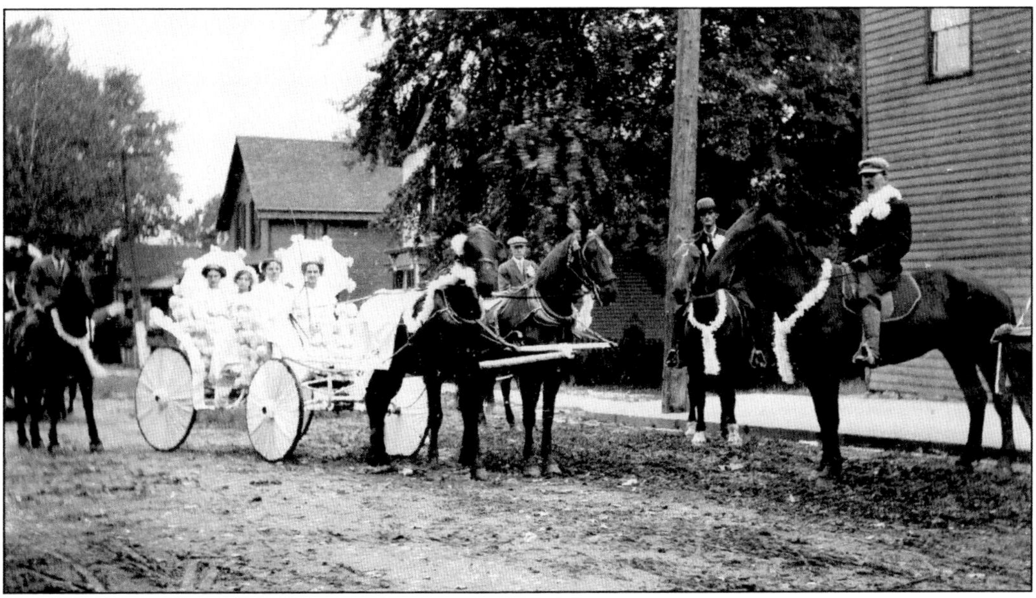

Frank Green's wife and children are seated in the decorated carriage with fancy parasols. Ed Chrisman (second from right) and Frank Green (right) are mounted on their horses. (Green collection.)

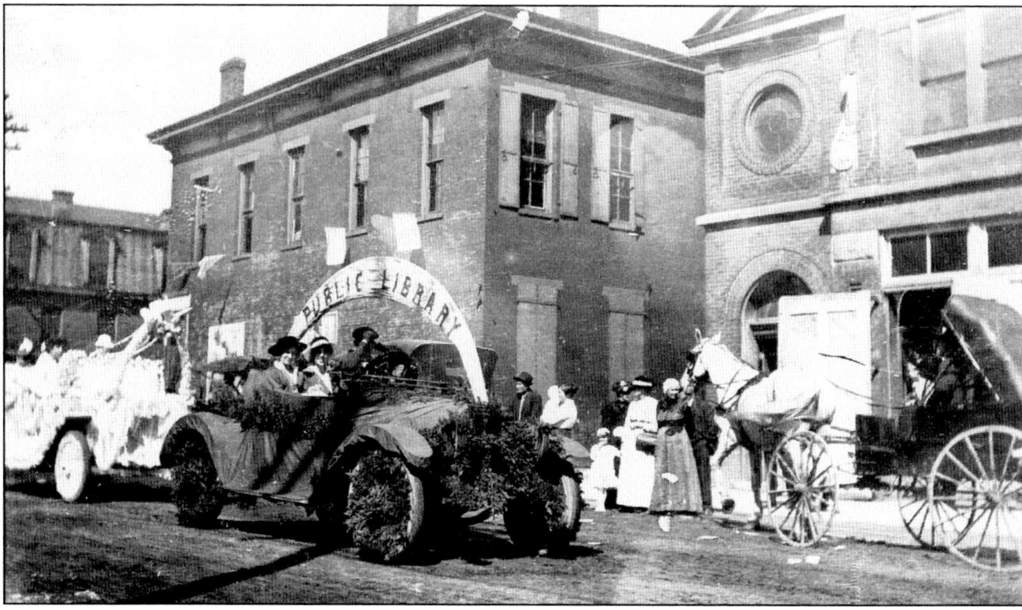

On October 11, 1913, an invitation was extended to every automobile owner to take part in an automobile parade. The Aurora Public Library decorated car was no doubt drumming up support for the new library. The Aurora Public Library was erected in 1914 on Second Street. The Ohio Valley Motor Club awarded cash prizes as high as $15 for the best decorated. F. H. Rieman won a first and J. C. Wright received a second for their entries. (Hizer collection.)

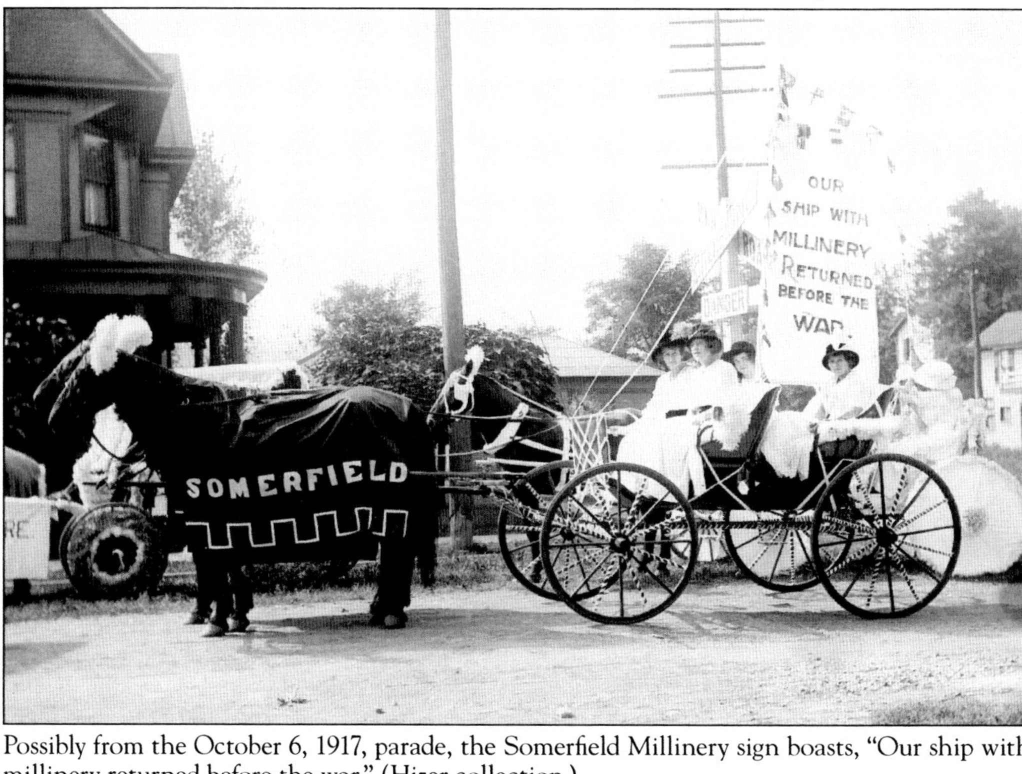

Possibly from the October 6, 1917, parade, the Somerfield Millinery sign boasts, "Our ship with millinery returned before the war." (Hizer collection.)

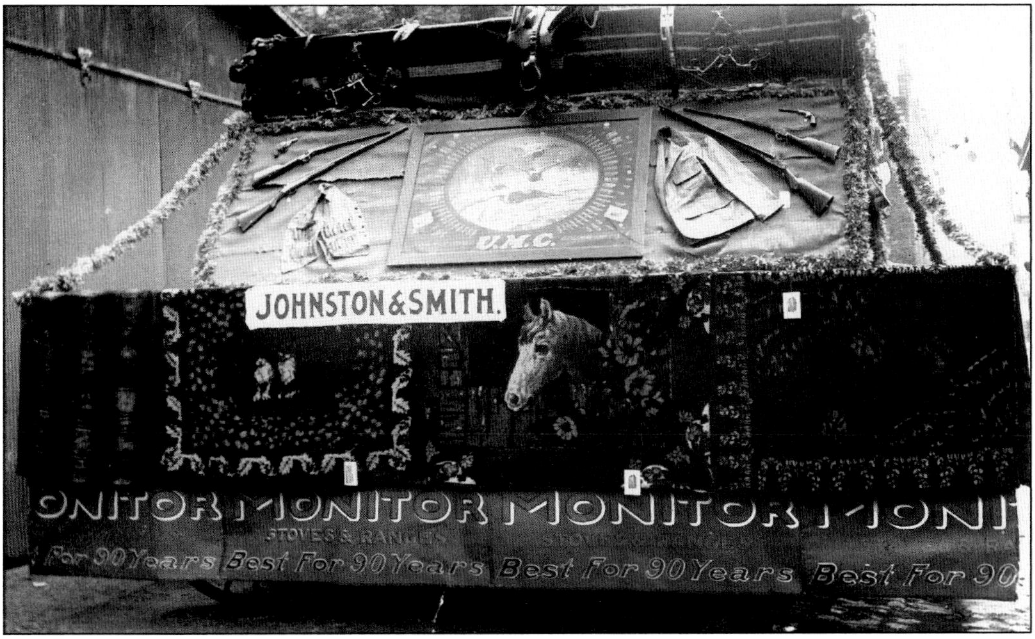

Johnston and Smith Furniture Store sold everything needed for a home, not only fine rugs and stoves, but also guns and hunting apparel. (Hizer collection.)

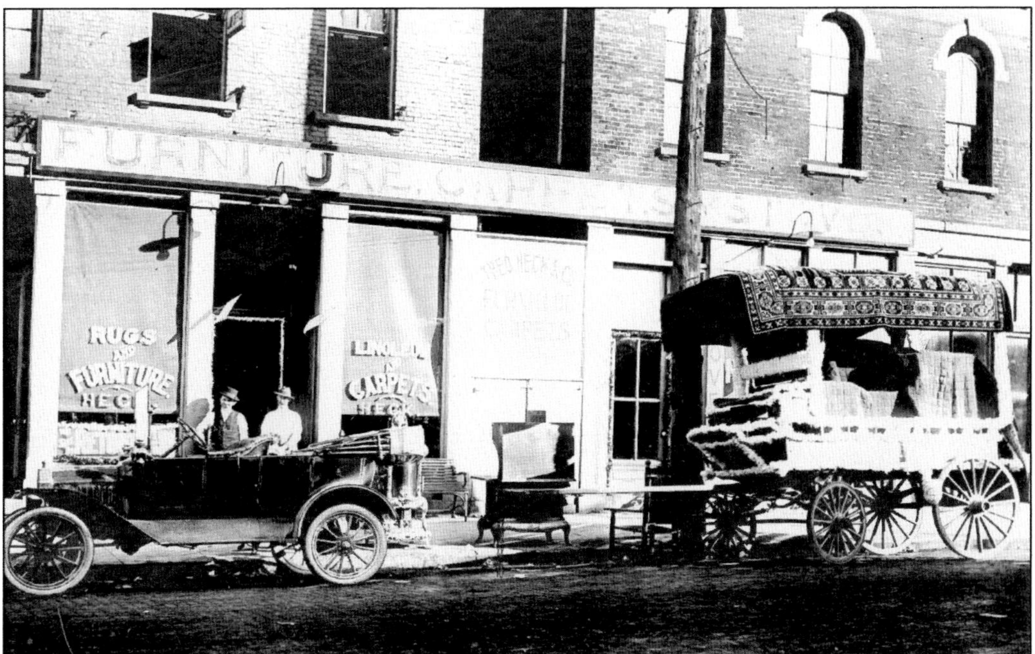

Frank Green and Andy Johnson, employees of the Heck Furniture Company (building on left) stand in front of what is now the Aurora City building. After advertising their wares in the parade, store owners would feature their displays on Main Street afterward. Exhibited in 1916 was Stedman's limestone pulverizer. The machine pulverized limestone from area creek beds and was used as fertilizer. (Green collection.)

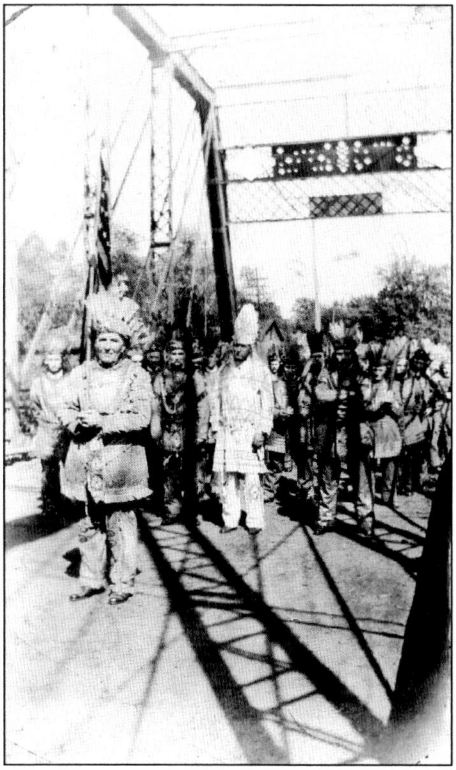

Both photographs are from the October 6, 1917, fair. Twenty young men attired as Colonials and American Indians pose on the George Street bridge. The Colonials represented "The Spirit of '76." (Aurora Lions Club collection.)

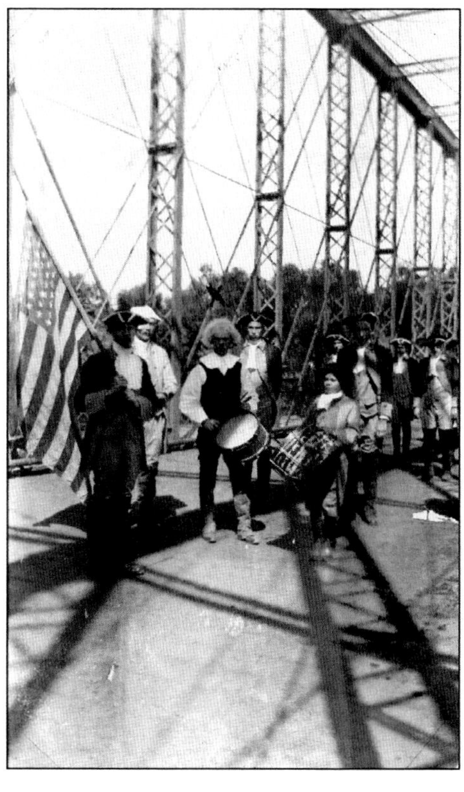

The gentleman with the snare drum is Gatch Baker. Gatch was born in nearby Wilmington and was the father of twin sons, Dr. Leslie Baker and Judge Lester Baker. Three periods in American history were represented in this 1917 parade. (Aurora Lions Club collection.)

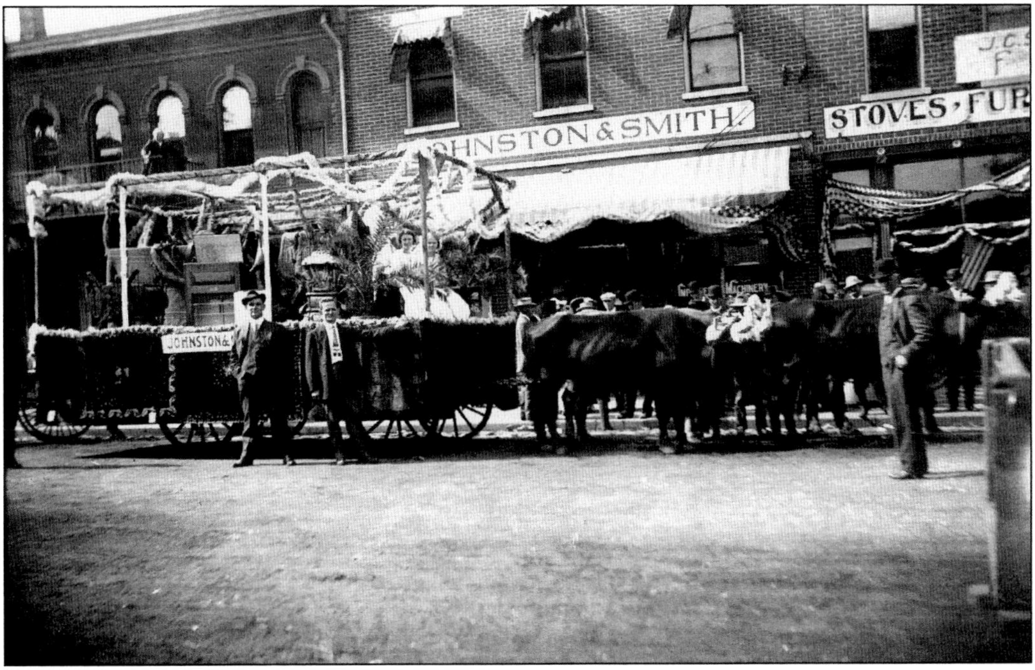

This 1912 business display by Johnston and Smith was a rare sight for the younger crowd. A double team of oxen with their horns gaily decorated pulled the wagon. Robert L. Johnston (standing far left) and his partner Smith (standing center) pose for a photograph in front of their furniture store located on Main Street. (C. Johnston collection.)

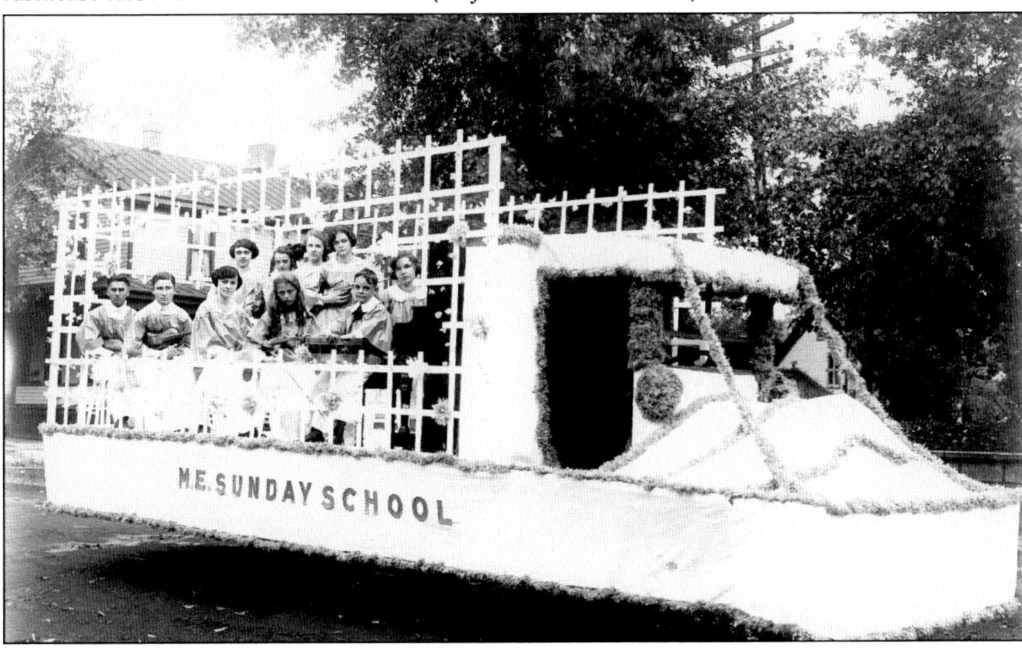

The Methodist Episcopal church Sunday school float is seen about 1923. Pictured from left to right are (seated) unidentified, Robert Jackson, Inez Baker, Ruth Morris, Clark Hiel, and Doris Smith at the piano; (standing) Ruth Hiel, Charlotte Hiel, Eleanor Stier, and Betty Matson. (Hizer collection.)

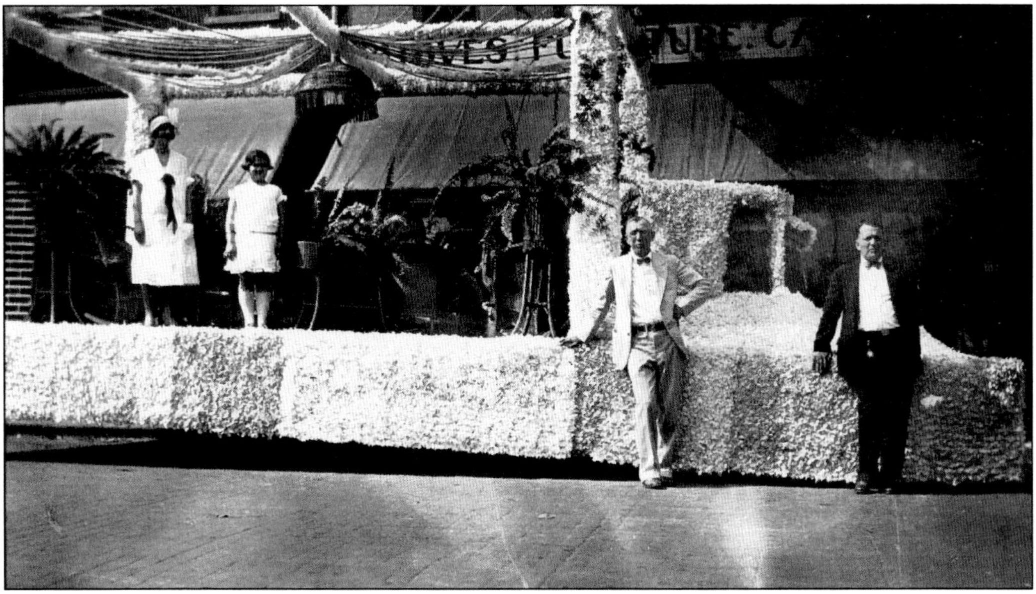

Seen here are the Schuler's display with nine-year-old Lucille Schuler and an unidentified woman standing on the float. Standing in front of the float are Frank and Raymond Schuler. The float displayed different types of reed furniture perhaps from the National Reed Fiber Furniture Company. (Craft collection.)

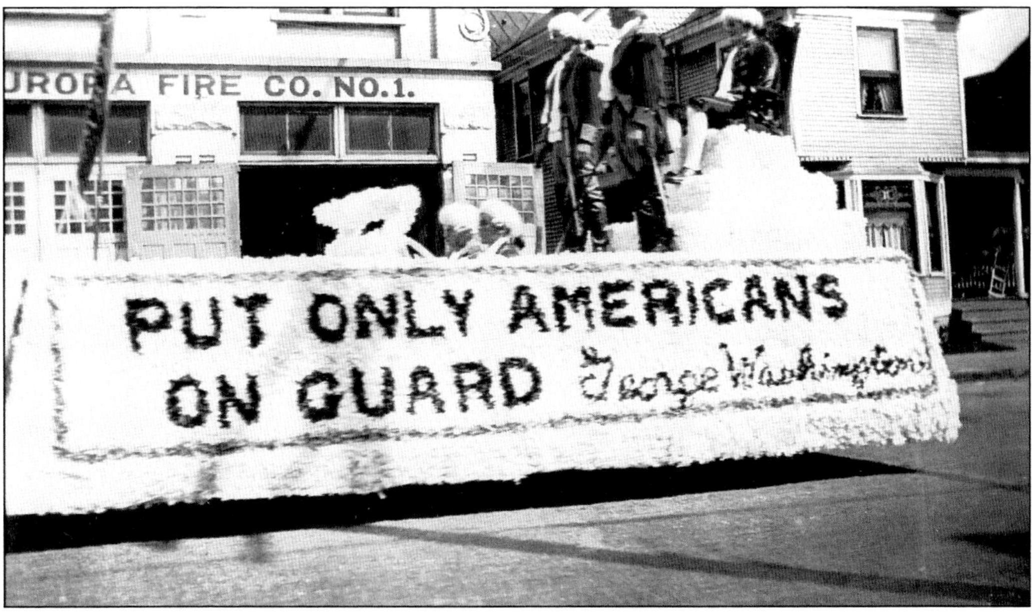

This early float is quoting George Washington, "Put Only Americans On Guard—George Washington." (Craft collection.)

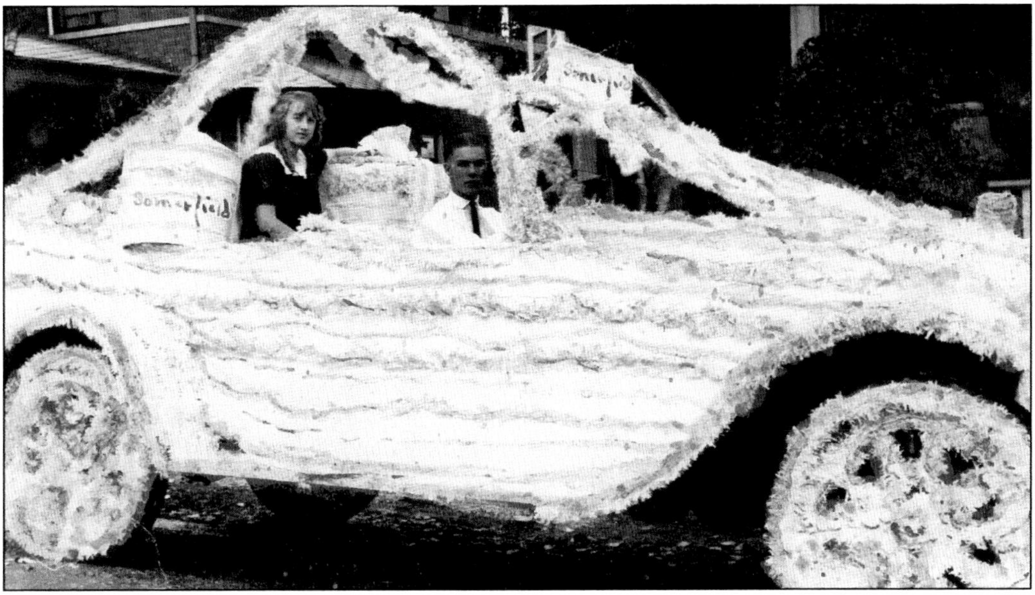

The Somerfield Millinery Store vehicle was a first-place winner on October 7, 1922. The millinery, located on Second Street, while in business, was always represented in the parade. (Aurora Lions Club collection.)

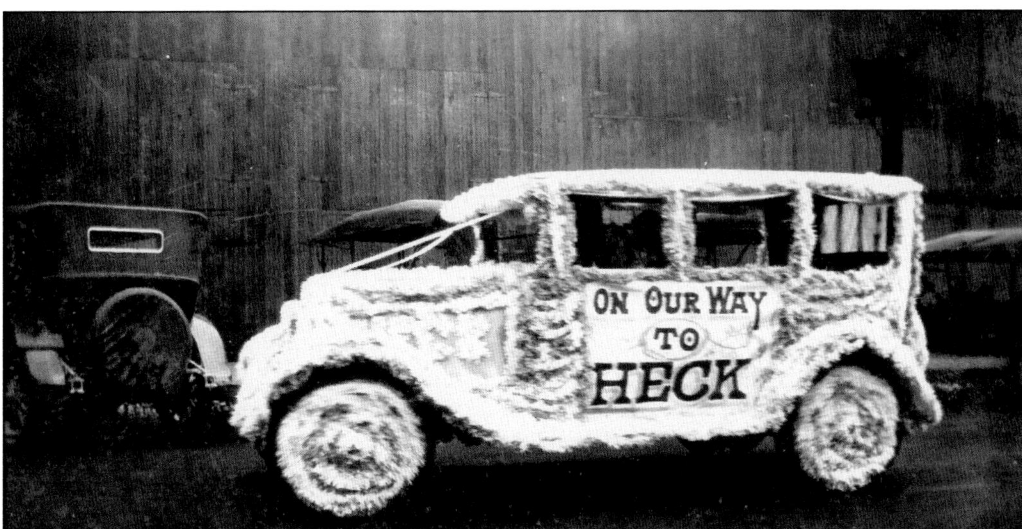

No longer displaying a horse-drawn float, Frank Green began driving decorated cars for his employer, Heck Furniture Store. (Green collection.)

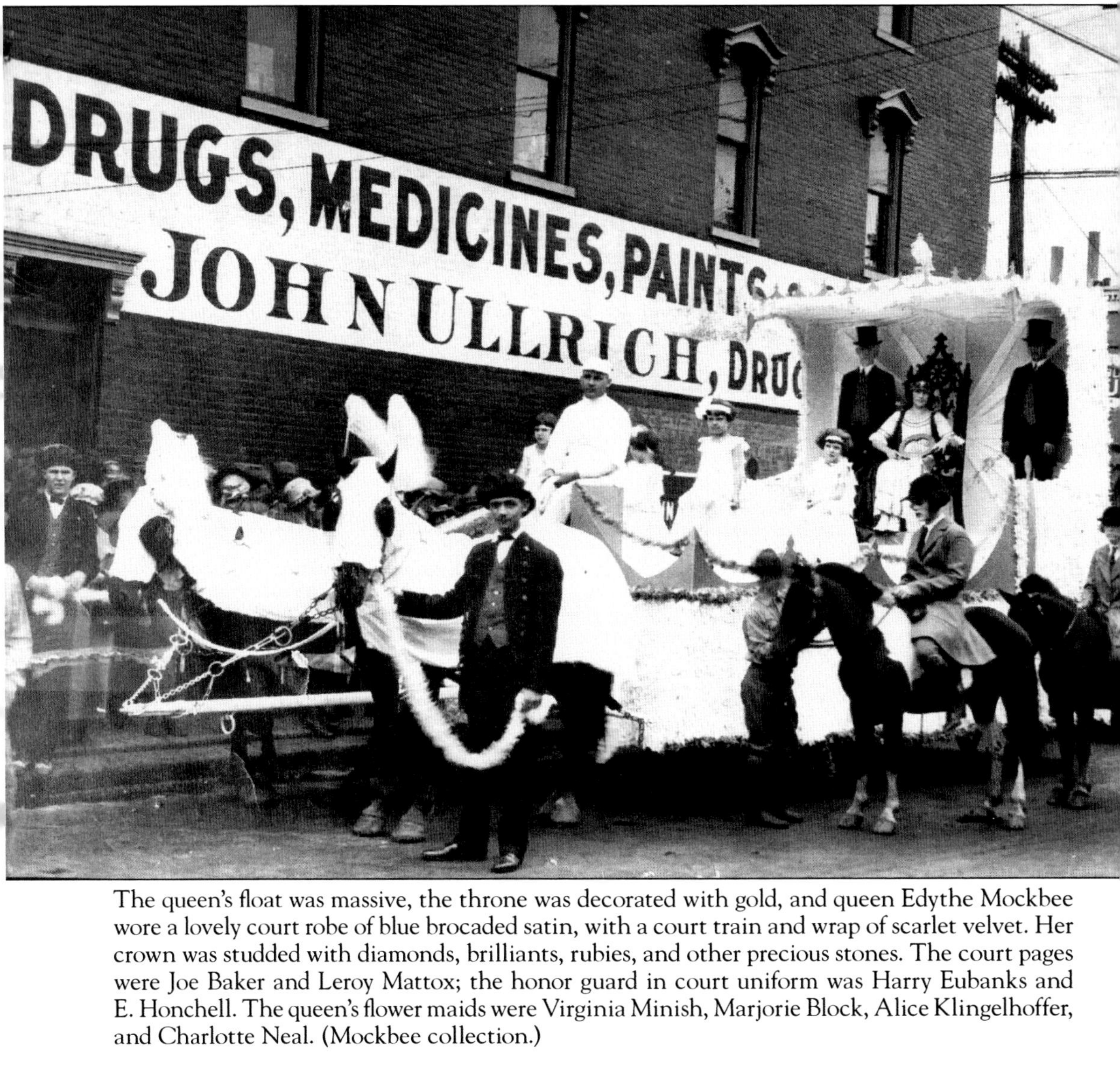

The queen's float was massive, the throne was decorated with gold, and queen Edythe Mockbee wore a lovely court robe of blue brocaded satin, with a court train and wrap of scarlet velvet. Her crown was studded with diamonds, brilliants, rubies, and other precious stones. The court pages were Joe Baker and Leroy Mattox; the honor guard in court uniform was Harry Eubanks and E. Honchell. The queen's flower maids were Virginia Minish, Marjorie Block, Alice Klingelhoffer, and Charlotte Neal. (Mockbee collection.)

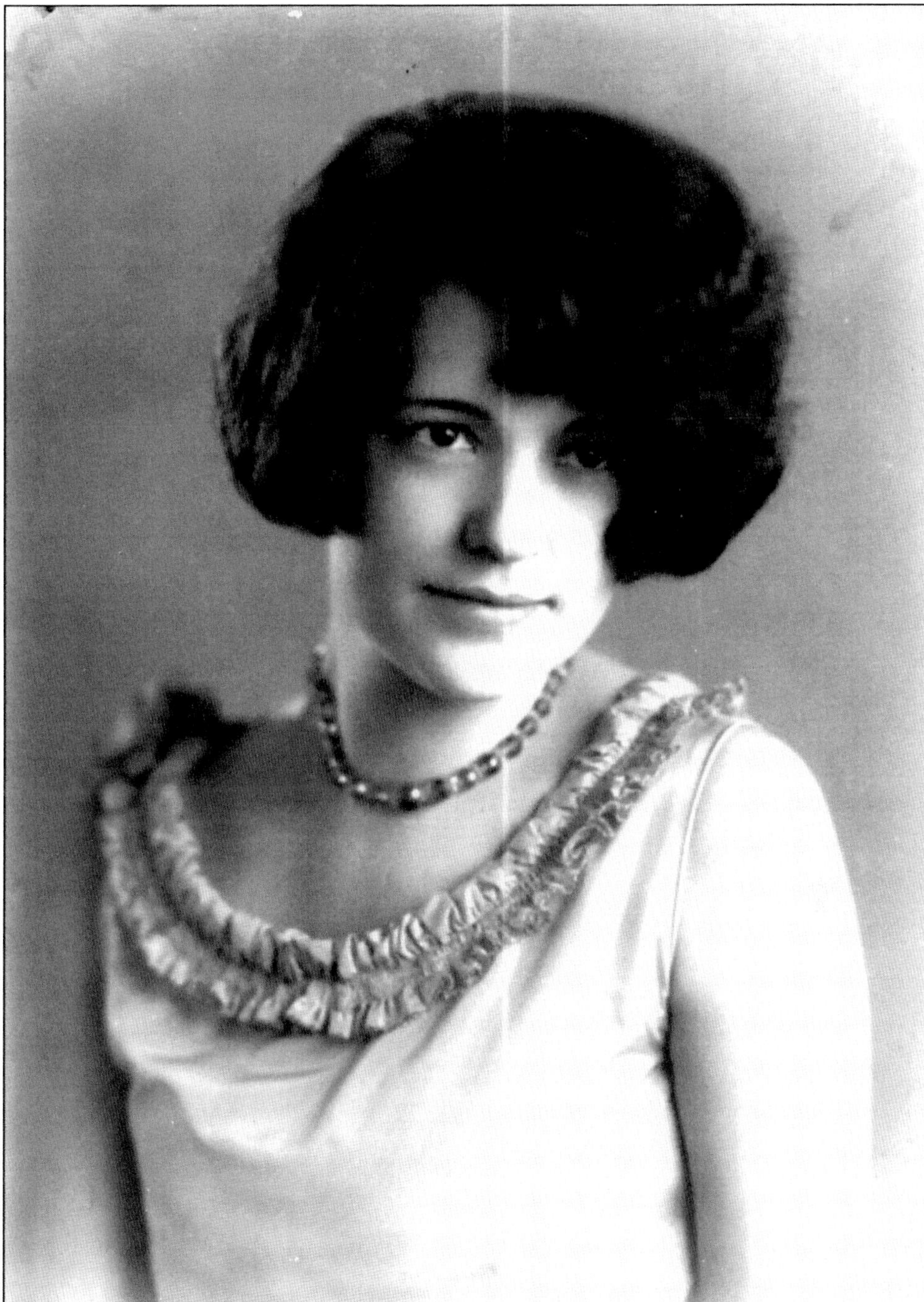

The 15th annual Aurora Farmers Fair of 1924 was the first fair to include a royalty competition. Edythe Mockbee, 22, was voted queen of the festival. The queen arrived from downriver on the speedboat the *Hoosier Girl*. Church bells announced her arrival, and she then was presented with the key to the city by Mayor Edward J. Libbert. Pearl Hueseman, Aurora, placed second and Evelyn Webber, Aurora, third. Mockbee would be the first and the last queen until 1951. (Mockbee collection.)

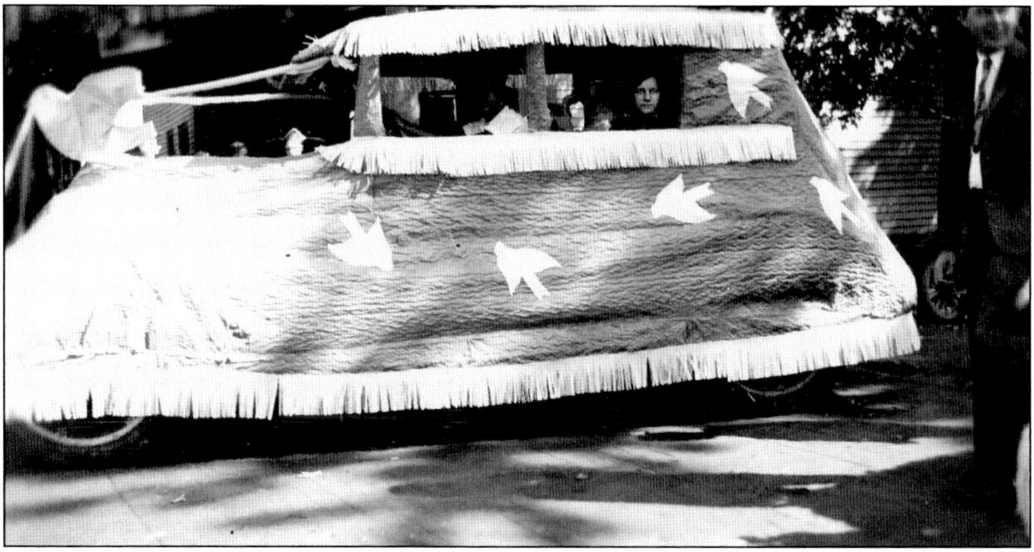

The First National Bank entered this white and silver trimmed car in 1925. Over the radiator were two white doves from which streamers were stretched back to little fairies. (Weaver collection.)

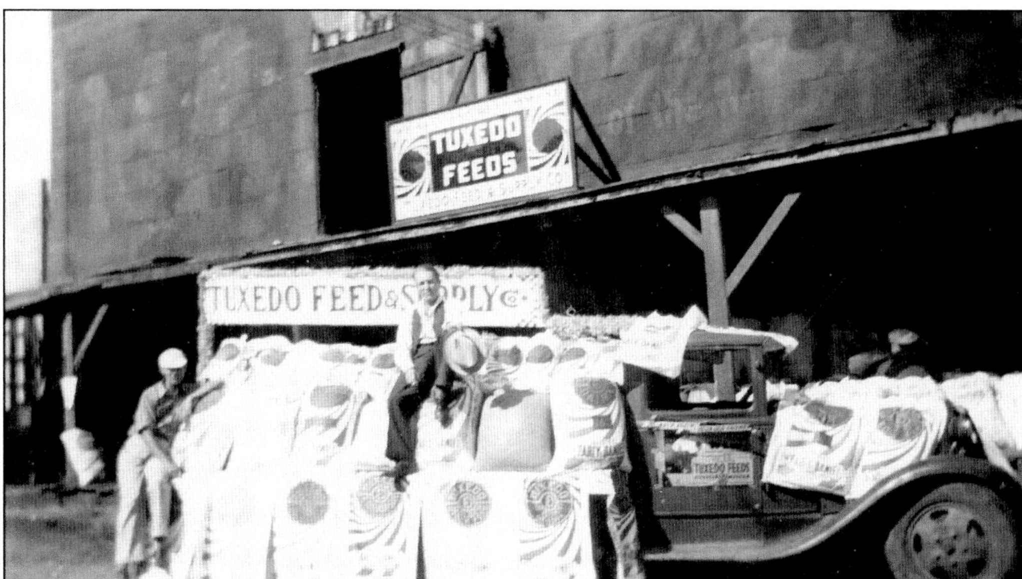

The first parade for the Aylor Feed Store was in 1935. Aylor sold Tuxedo Feeds. (Aylor collection.)

Two

FROM 1930 TO 1949

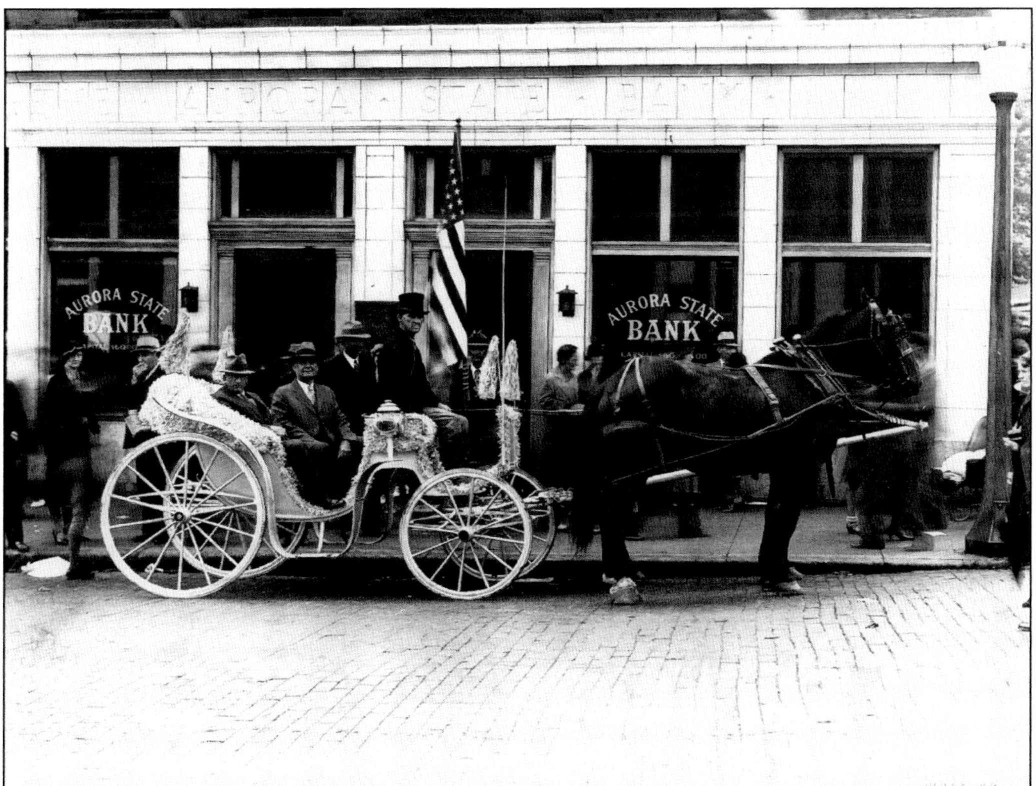

Rain did not dampen the spirit of the 15,000 visitors celebrating the fair's 25th anniversary on October 7, 1933. Riding in this horse-drawn carriage is Maurice C. "Boss" Johnston, president of the 1933 Farmers Fair Association, and C. B. Wilson, both originators of the first farmers fair. The carriage shown was loaned by Mrs. H. W. Smith and the horses were loaned by Mrs. Martin J. Givan. (Ellinghausen collection.)

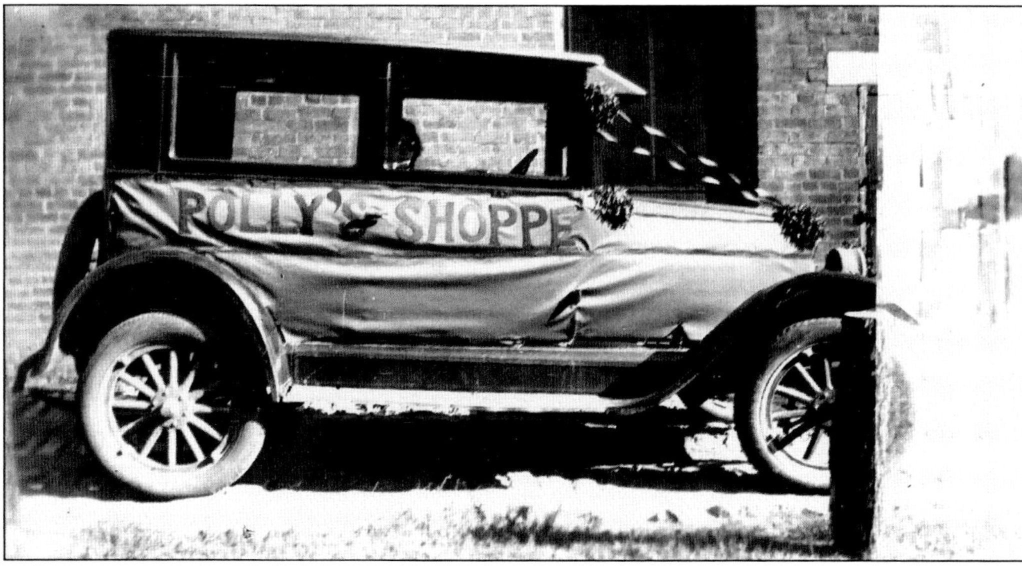

Polly's Shoppe entered this decorated automobile. Polly's Shoppe was a card and gift store. (Weaver collection.)

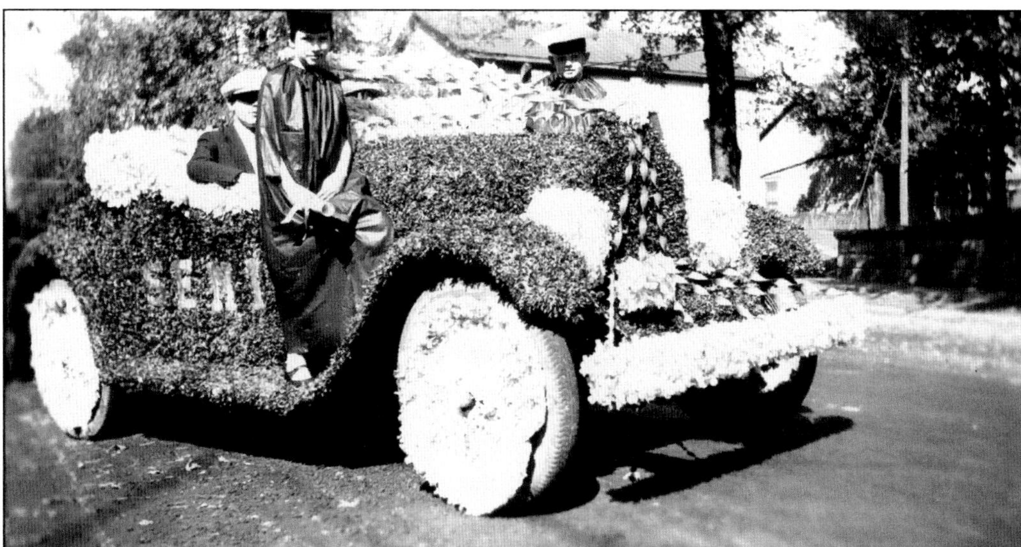

Aurora High School seniors rode on the sideboards of this decorated automobile. (Weaver collection.)

Taylor Brothers Company first brought the Model-T "Bucking Ford" to Aurora in 1935. In the back seat is Frank Taylor. Dutch Boyles would drive and cause the bucking motion with clutches. The driver handled the left brake for sharp left turns. Passenger Louie Wells worked the right brake and fired a sawed-off shotgun located in the floorboard. This car evolved into "Peggy the flying red horse." (J. F. Ullrich collection.)

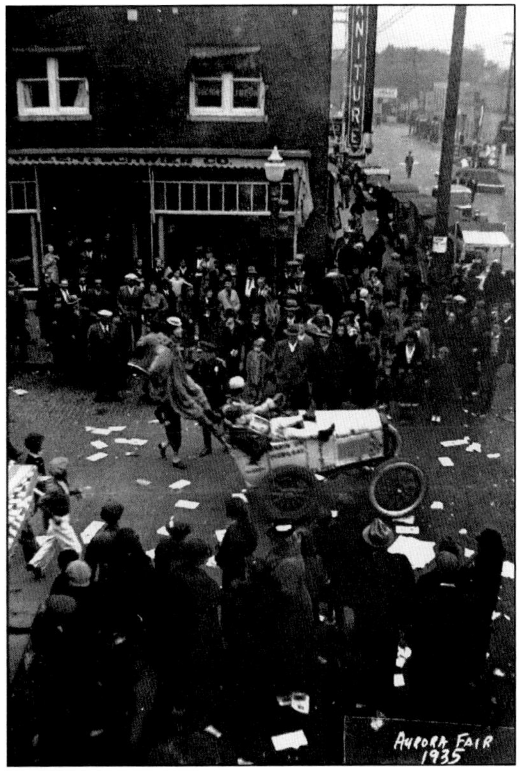

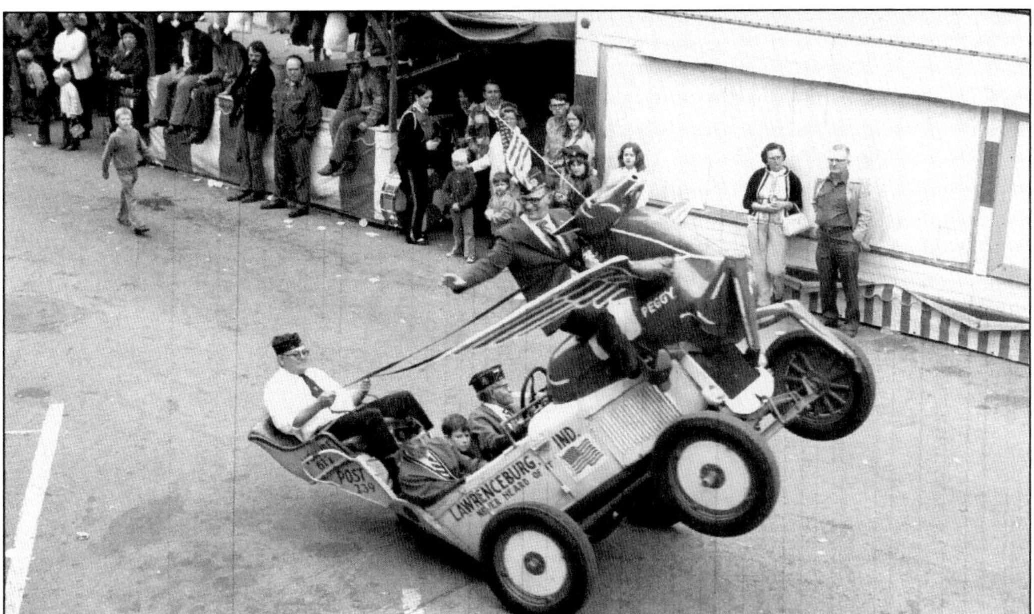

The Taylor Brothers flying red horse, nicknamed Peggy, was named for the Pegasus of mythology and the logo for Mobile Oil. The Boyles brothers rebuilt the Bucking Ford body. Louis Wells built the horse on front. Frank Taylor built the wings. A gear moved the wings up and down. Peggy is retired and on permanent display at the Lawrenceburg American Legion. (J. F. Ullrich collection.)

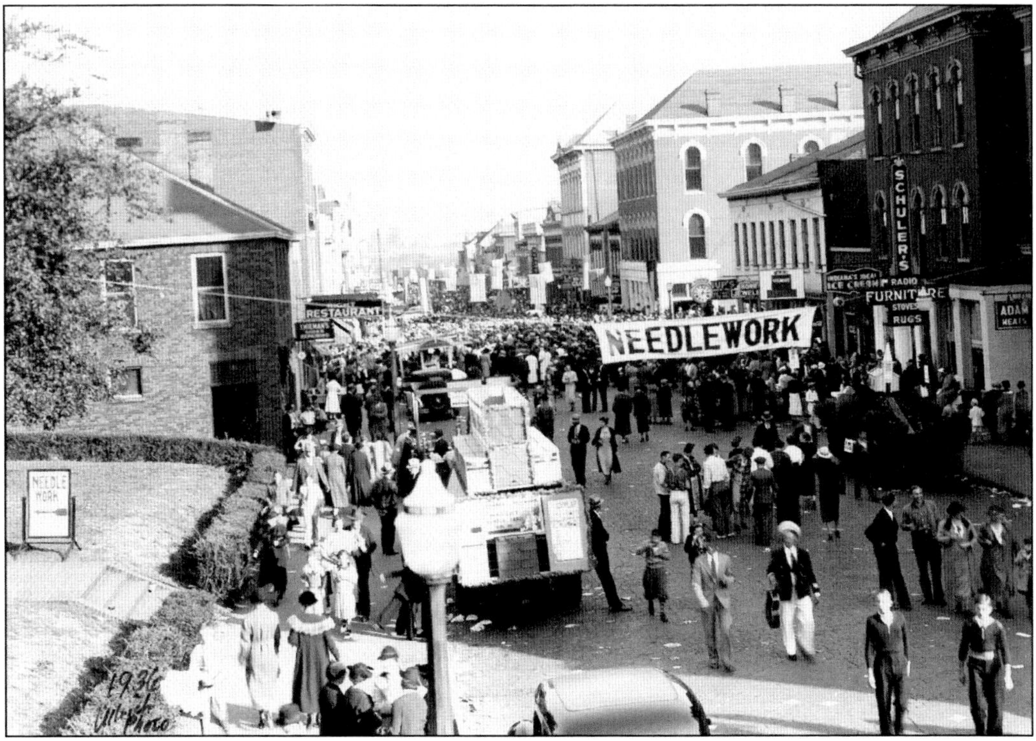

In 1936, the needlework was displayed in the public library. The crowd inspects various floats displayed on Second Street after the parade. (J. F. Ullrich collection.)

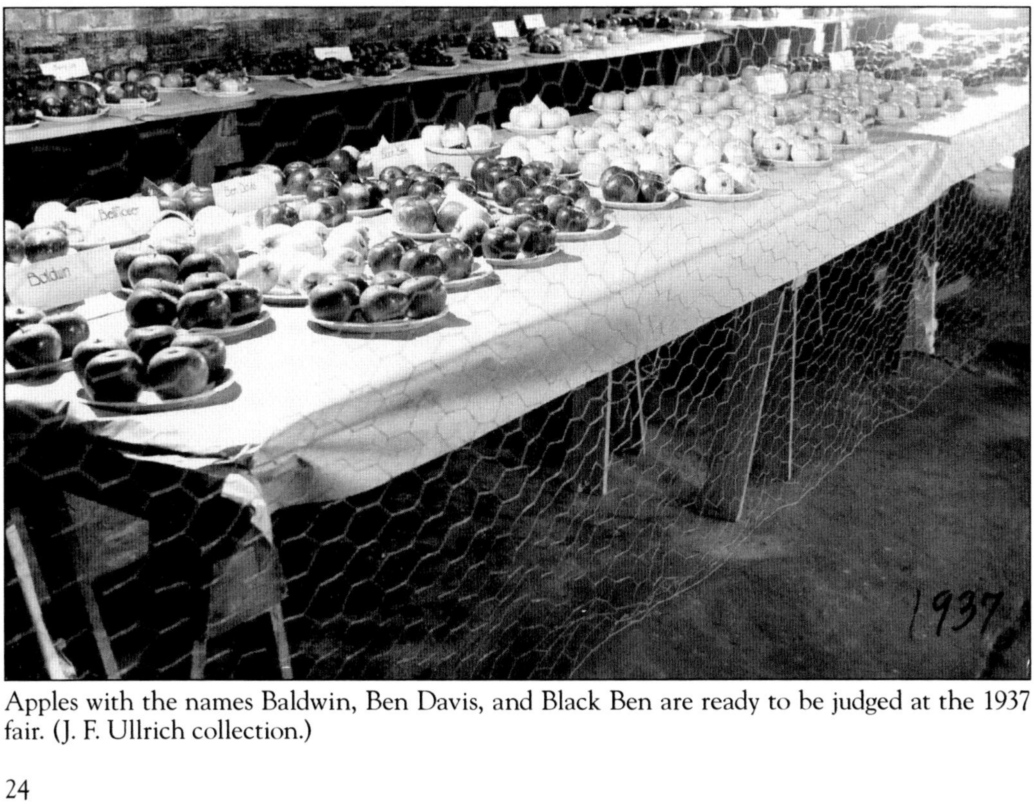

Apples with the names Baldwin, Ben Davis, and Black Ben are ready to be judged at the 1937 fair. (J. F. Ullrich collection.)

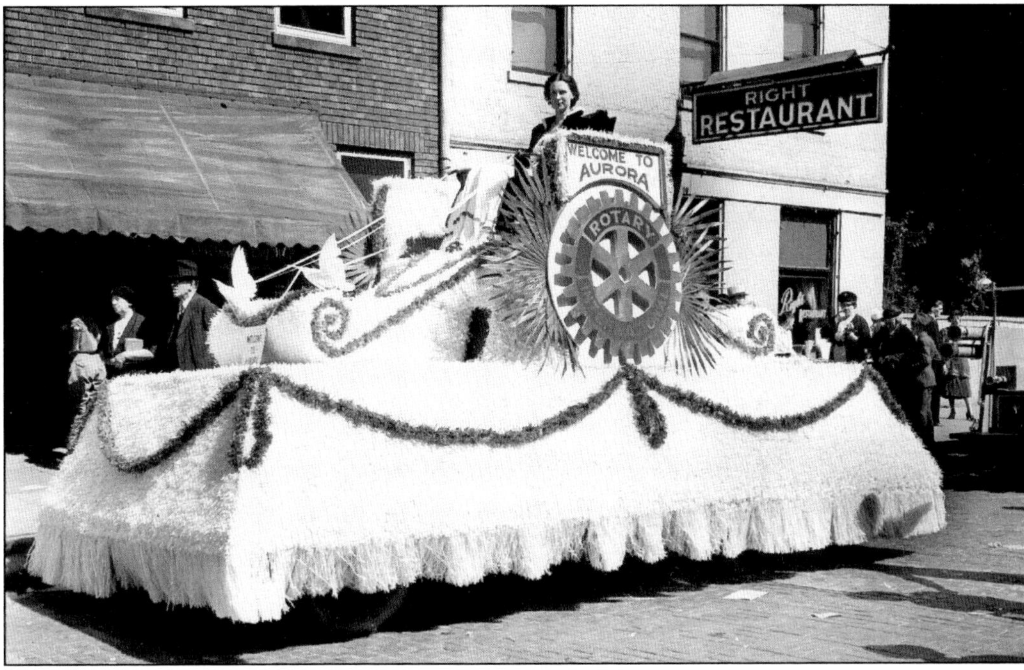

This beautiful 1938 Aurora Rotary Club float carried the greeting "Welcome to Aurora." Happy to have the flood of 1937 behind them, and the town cleaned up, the 30th annual farmers fair celebrated a new beginning for the flood-ravaged town. The young lady seated is unidentified. (J. F. Ullrich collection.)

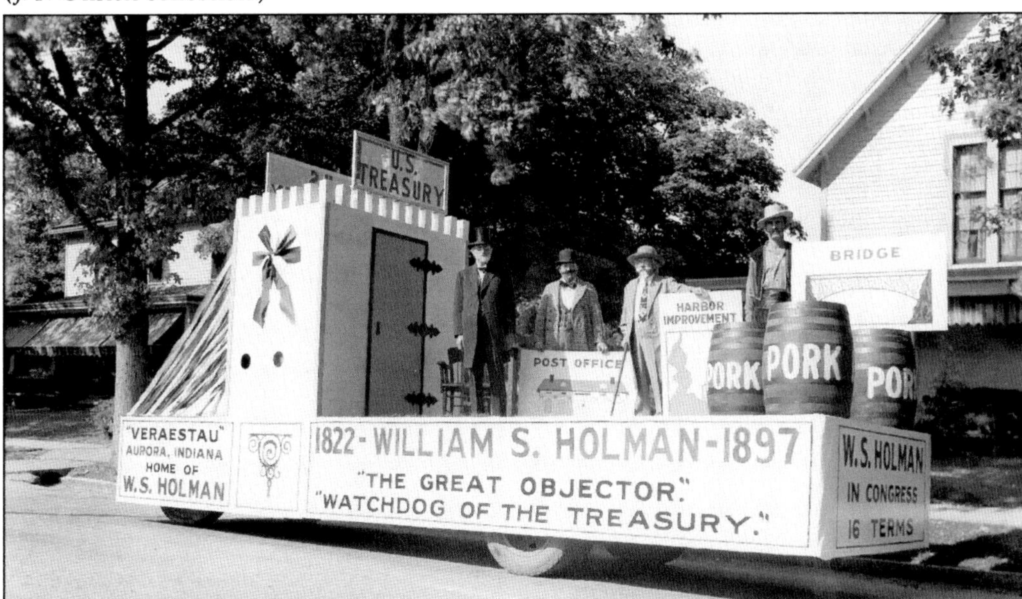

This 1938 float celebrates William S. Holman, born in 1822 in a home named Veraestau, overlooking Aurora. Known as the "Watchdog of the United States Treasury," Holman had a long political career. He was a member of the Indiana House of Representatives from 1851 to 1852 and elected to the U.S. House of Representatives in 1859. Over the following 38 years, Holman served in all but four Congresses. (J. F. Ullrich collection.)

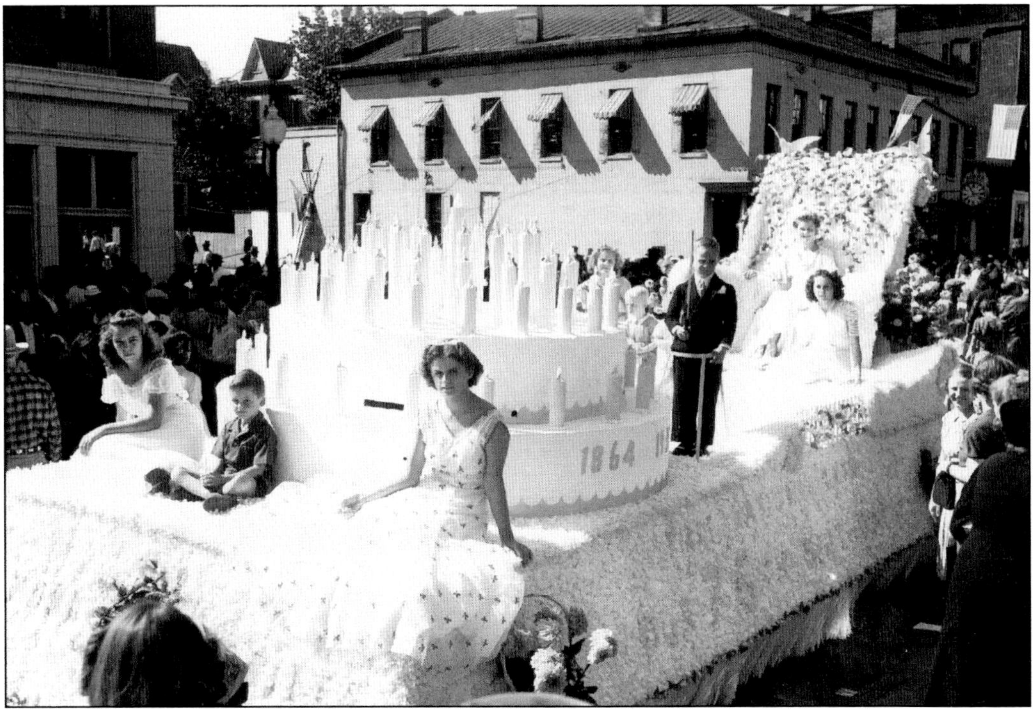

First prize in the 1939 fair was the diamond jubilee float of the First National Bank. Second prize for the most beautiful float in the parade went to Teaney Brothers. (J. F. Ullrich collection.)

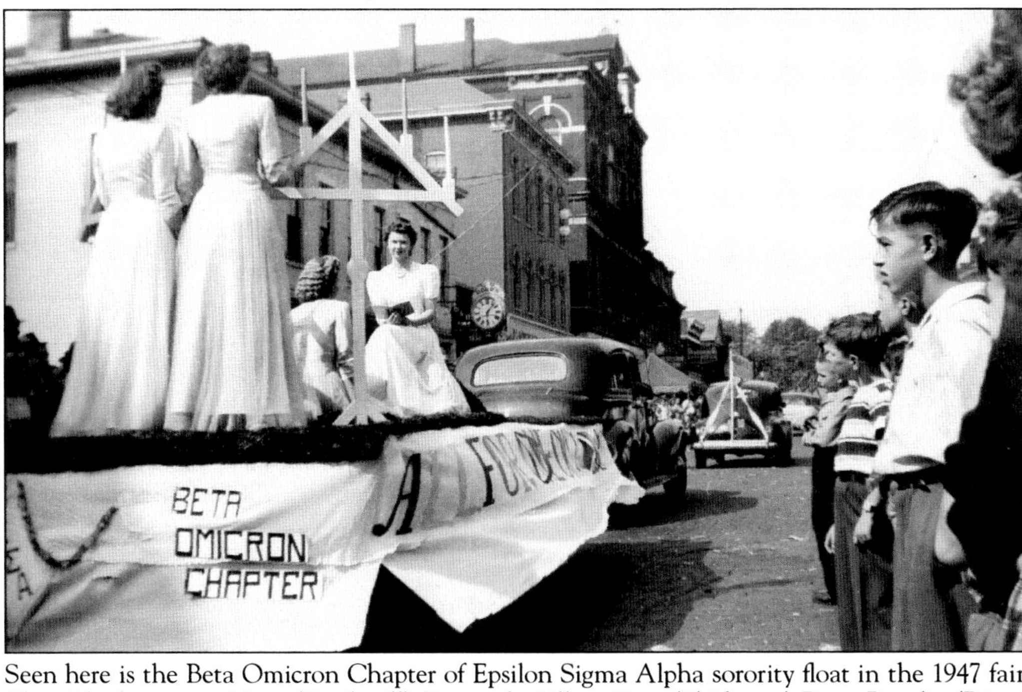

Seen here is the Beta Omicron Chapter of Epsilon Sigma Alpha sorority float in the 1947 fair. The girls shown are Mary (Satchwill) Hancock, Lillian Faye (Skidmore) Ross, Rosalie (Riggs) Kippler (kneeling), and Georgia Lee (Klingelhoffer) Dewers (seated). Alma Leah (Witte) Trennepohl was the chairman for this float. (Trennepohl collection.)

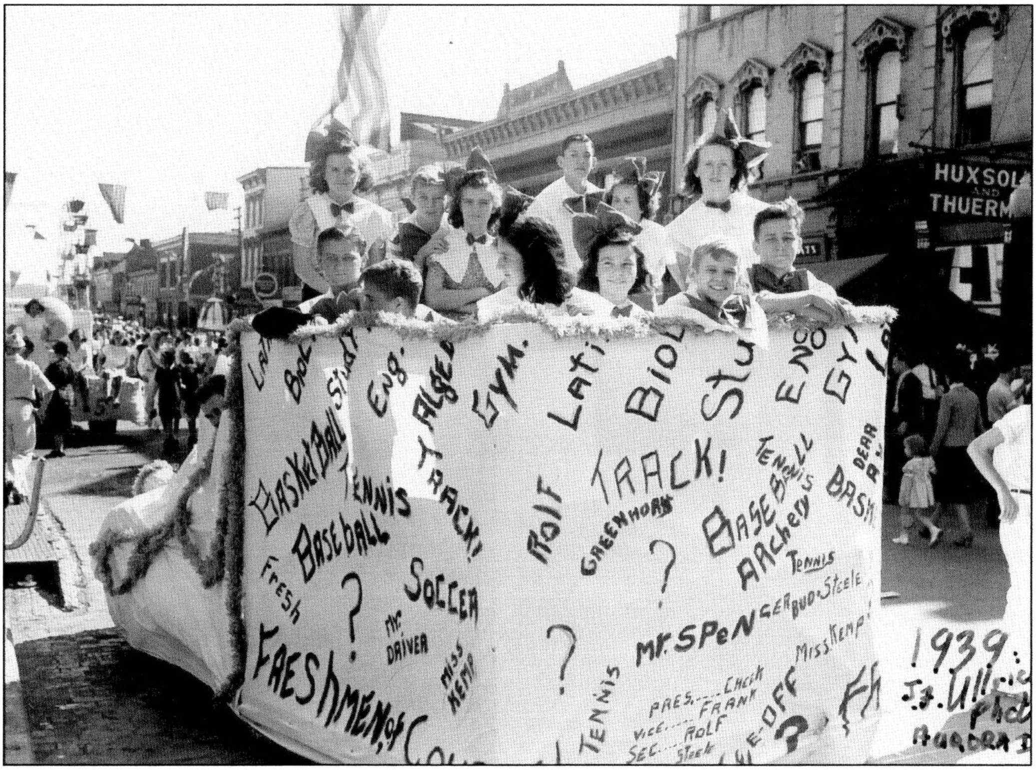

The 1939 Aurora freshman class float is seen here. (J .F. Ullrich collection.)

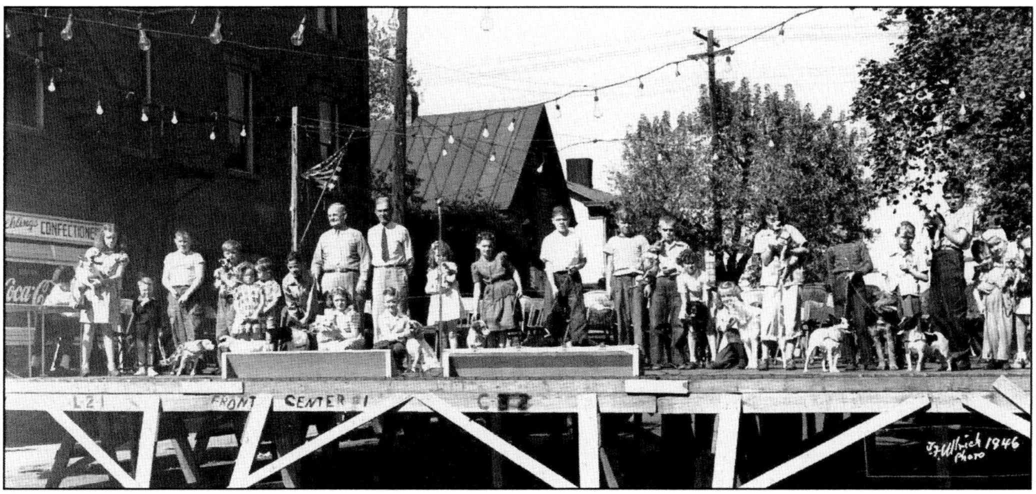

The 1946 pet parade was presented on Friday afternoon. Contestants pose onstage as the best dogs are judged. John Sutton won first place for the prettiest dog. Dickie Fussnecker's dog received a first place for the ugliest. (J. F. Ullrich collection.)

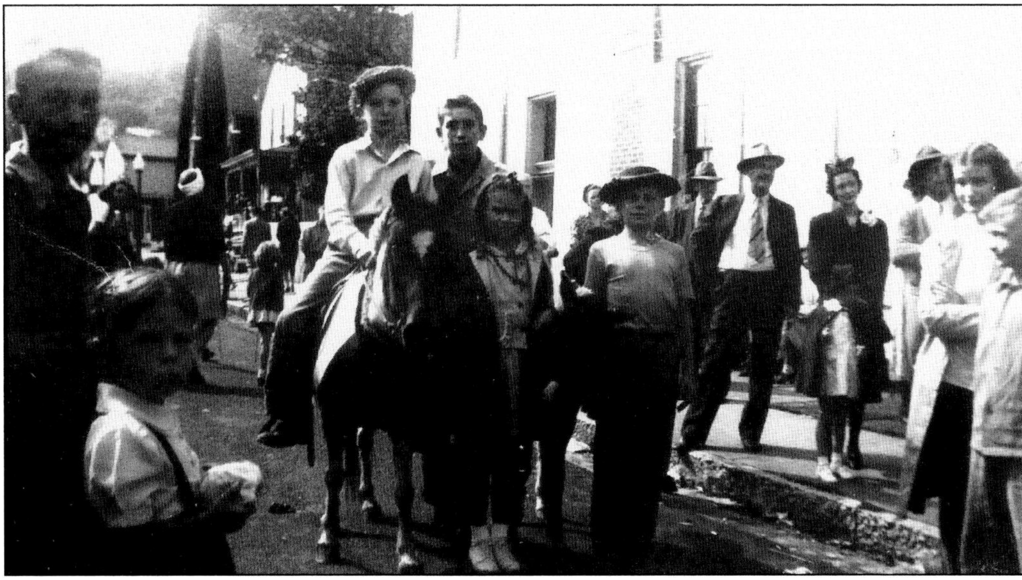

The Largent children are with their ponies in the Friday afternoon pet parade. From left to right are J. D. Largent on his pony Daisy June, Jim Cole, Nancy Anne Smith, and Tommy Largent holding Daisy's colt. The Largent children sent this photograph to their father, who was serving in the U.S. Army in World War II. Their father cherished this photograph and carried it with him while in Berlin, Germany, and crossing the ocean twice. (Largent collection.)

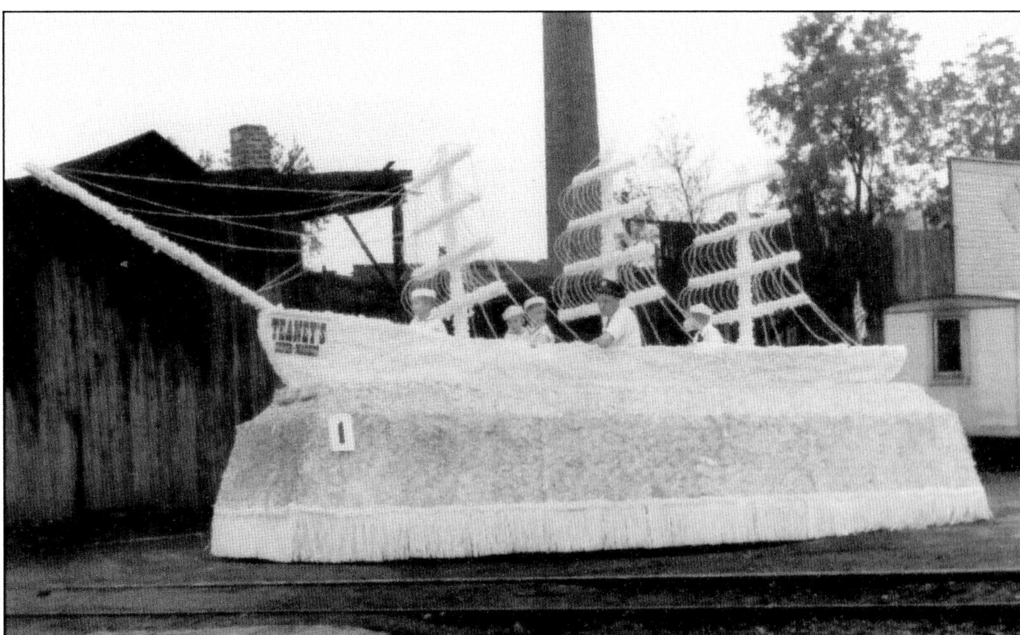

This 1941 float by Teaney Grocery Store won prettiest float. Sitting in the ship is a young Dan Teaney and Carl Petty. (Petty collection.)

Three

From 1950 to 1969

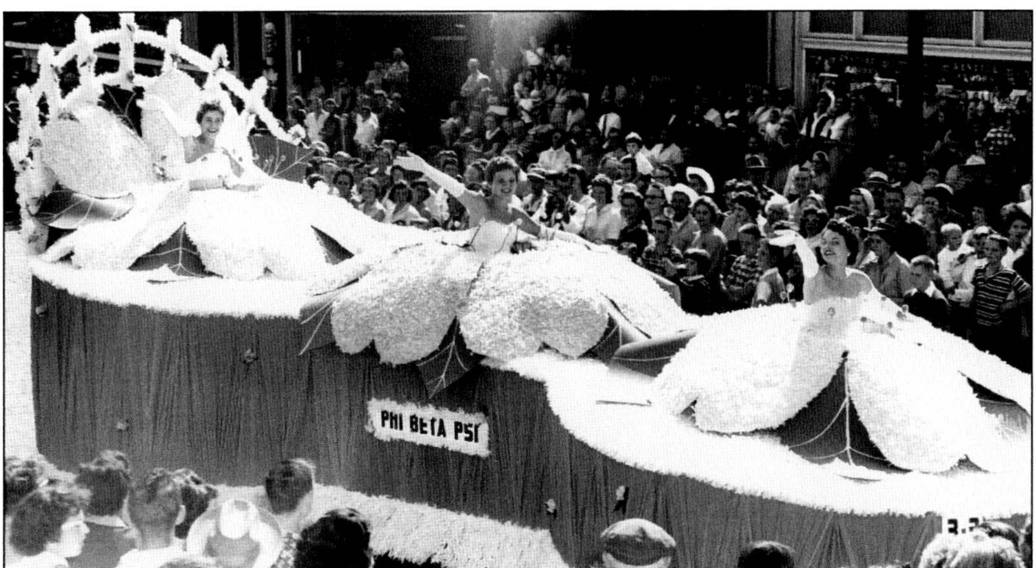

The rose is the flower of the Phi Beta Psi sorority, and on this float are three roses centered with pretty girls depicting the Greek letters of the sorority. Shirley (Peters) Meyer represented phi. Sue Downey was beta, and Mary Ann Teaney represented psi. The float measured 120 square feet of chip-celo sheeting with 200 yards of festooning and moved along by its own power. (J. F. Ullrich collection.)

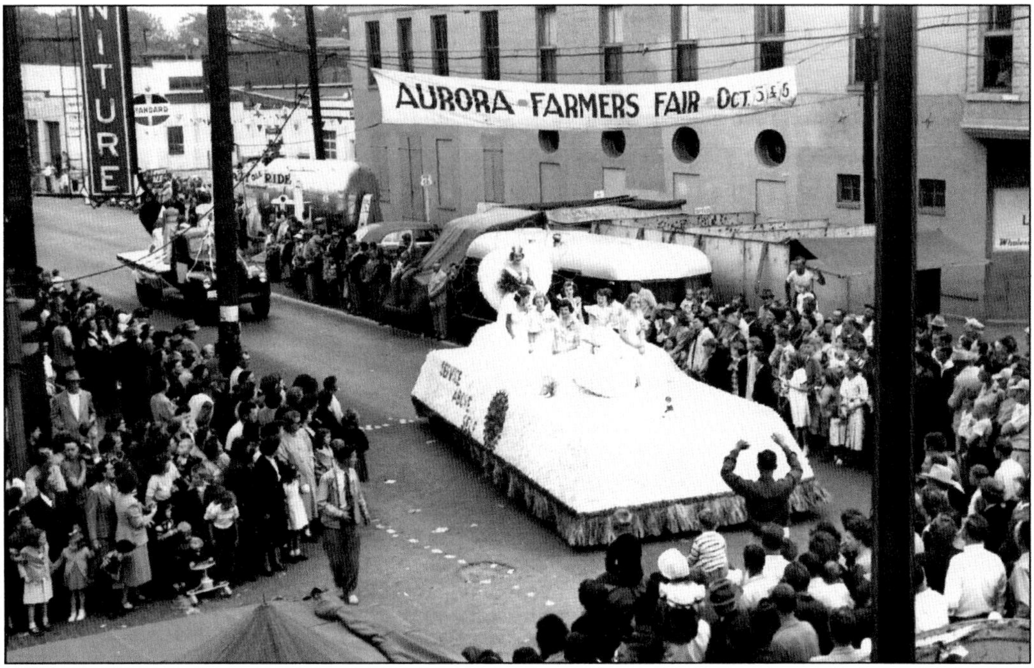

Twenty-seven years passed before another group of candidates would vie for the title of queen of the farmers fair. In 1951, Cecelia Dennis of Lawrenceburg won that title. Other contestants were Jerry Orme, Moores Hill; Betty Gordon, Dillsboro; Anita Andwin, Guilford; Mary Jane Gibson, Rising Sun; Margie Yauger, Bright; and Audrey Storey, Aurora. The Aurora Rotary Club sponsored the queen's float. (S. Ullrich collection.)

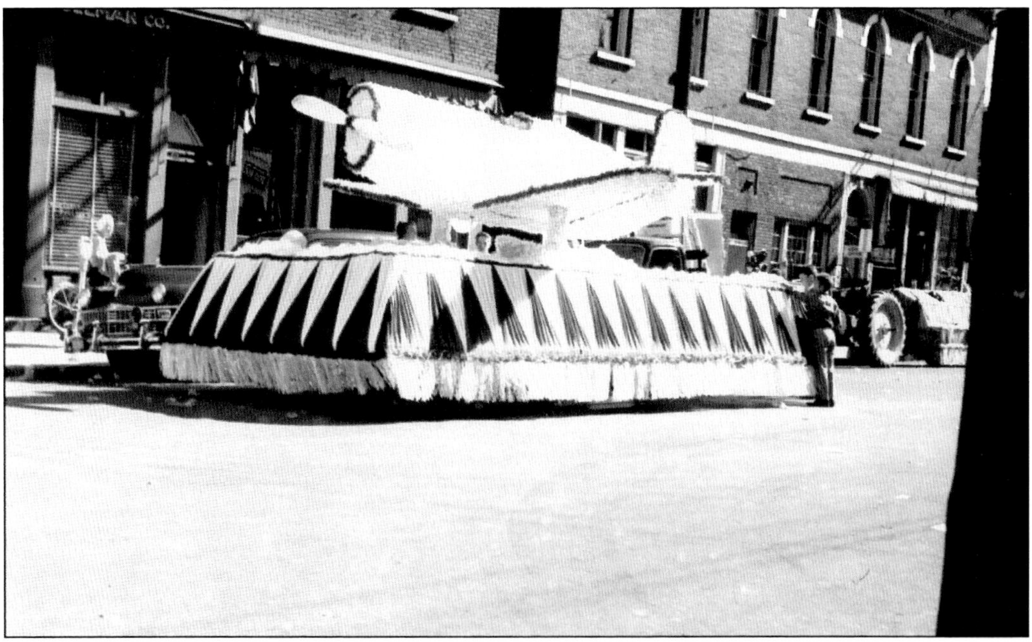

A very detailed white airplane sets on Main Street in front of the Holthause-Rullman Funeral Home for display. The 1949 Chrysler hearse and limousine is parked just behind the float and is still owned by Roger Rullman. (S. Ullrich collection.)

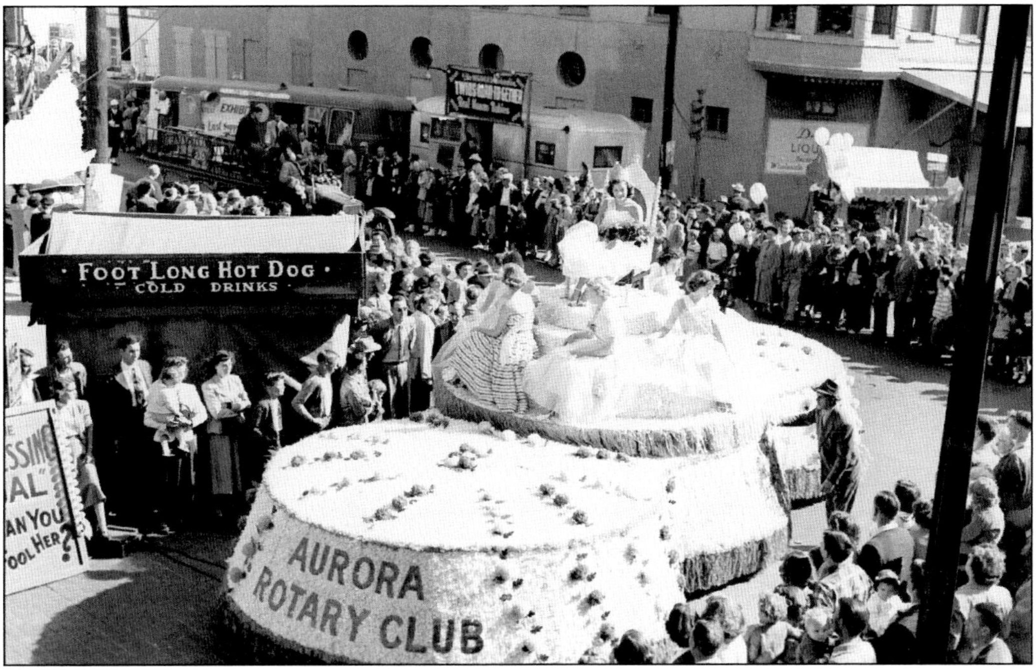

Jean Seibert, of Lawrenceburg, was crowned the 1952 queen. Attendants were Nancy Barrott, Aurora; Joan Beck, Bright; Helen Droege, Dillsboro; Joyce Hutchins, Moores Hill; Mary Jane Gibson, Rising Sun; and Ruth Zimmer, Guilford. Built on a Willies Jeep, the queen's chair was stationary. For the rotating bottom tier, Ralph Evans made an iron gear, together with a jackshaft and pinion that operated by a separate gasoline engine. (J. F. Ullrich collection.)

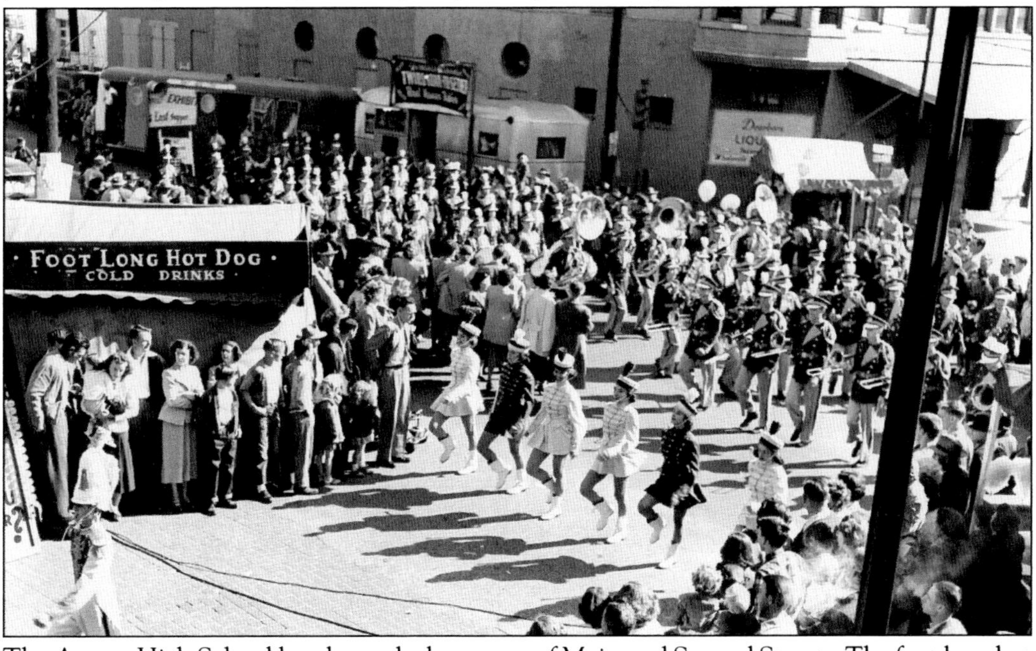

The Aurora High School band rounds the corner of Main and Second Streets. The foot-long hot dog stand in the background was a fixture on that corner for many years. Each year the vendors set up on the opposite side of the street from the year before. (J. F. Ullrich collection.)

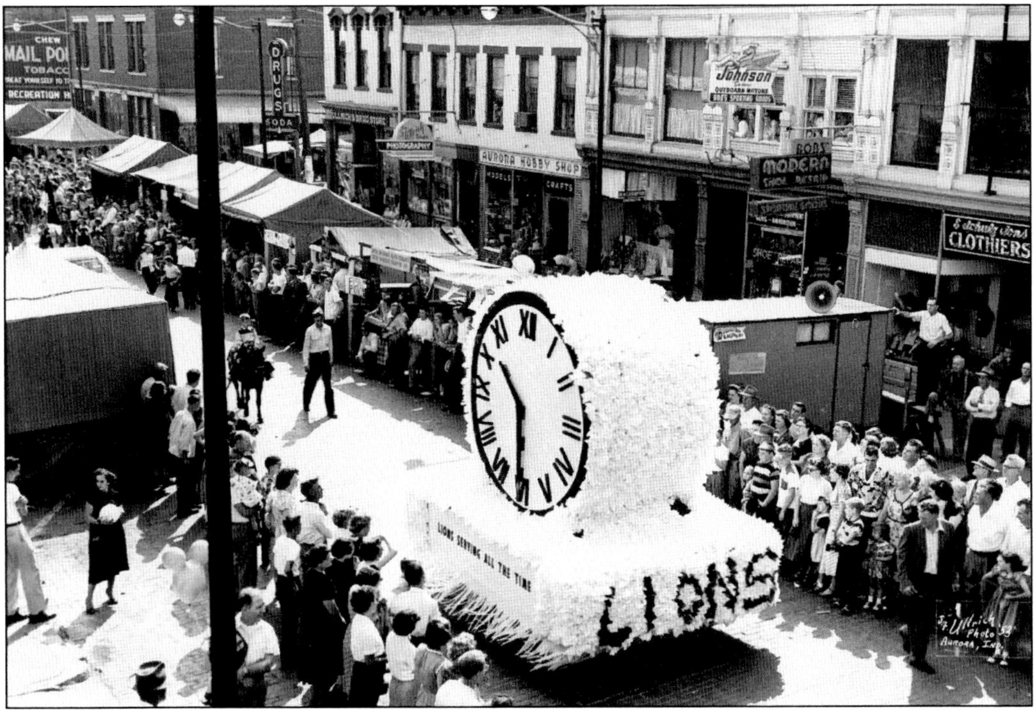

The Lions Club 1953 float theme was "Lions serving all the time." Notice the clock has a hidden driver and seems to float down the street on its own power. (J. F. Ullrich collection.)

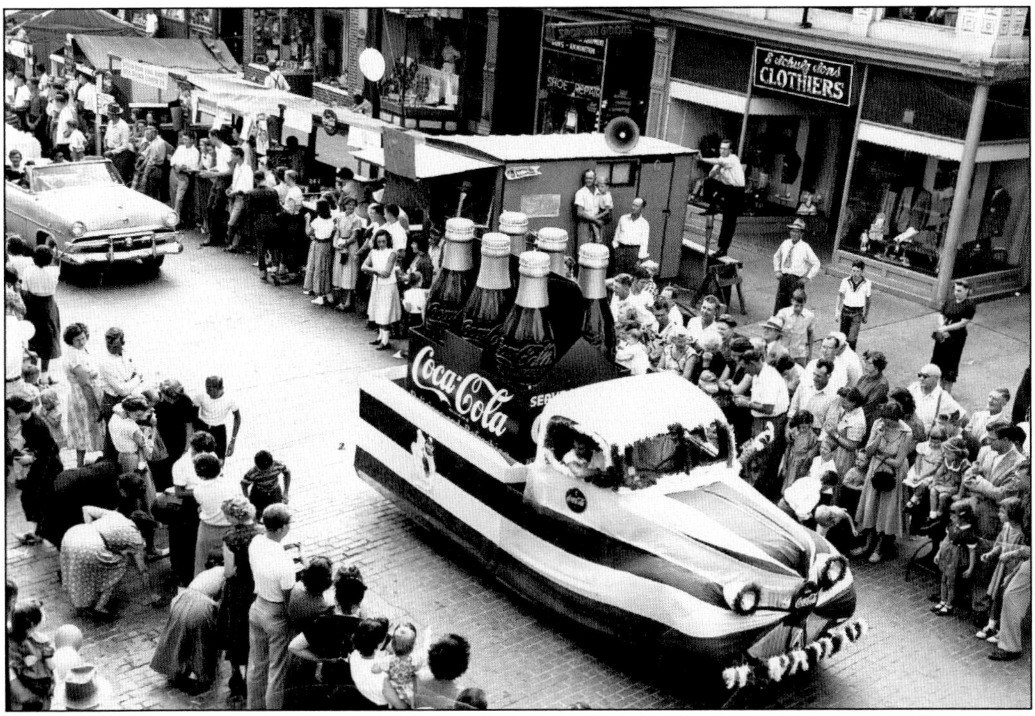

Children catch candy from a 1953 decorated truck representing the Aurora Coca-Cola distributing company. (J. F. Ullrich collection.)

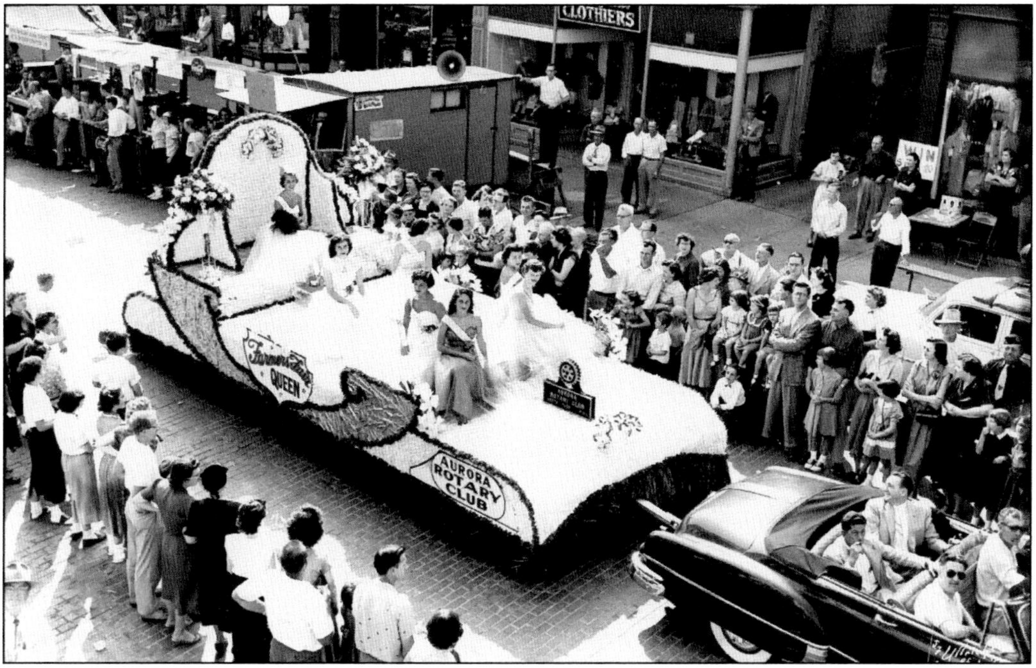

This Aurora Rotary Club float carries 1953 queen, Patti Greathouse, Aurora. Greathouse is shown on her throne while contestants on the left are Jackie Phillips, Guilford; Sally Frankel, Lawrenceburg; and Joyce Felix, Moores Hill. Contestants on the right side are LaVerne Jeffries, Bright; Garnetta Huber, Dillsboro; and Leitha Busse, Rising Sun. Ben Withered and Doc Jackson are seated in the back seat of the convertible. (J. F. Ullrich collection.)

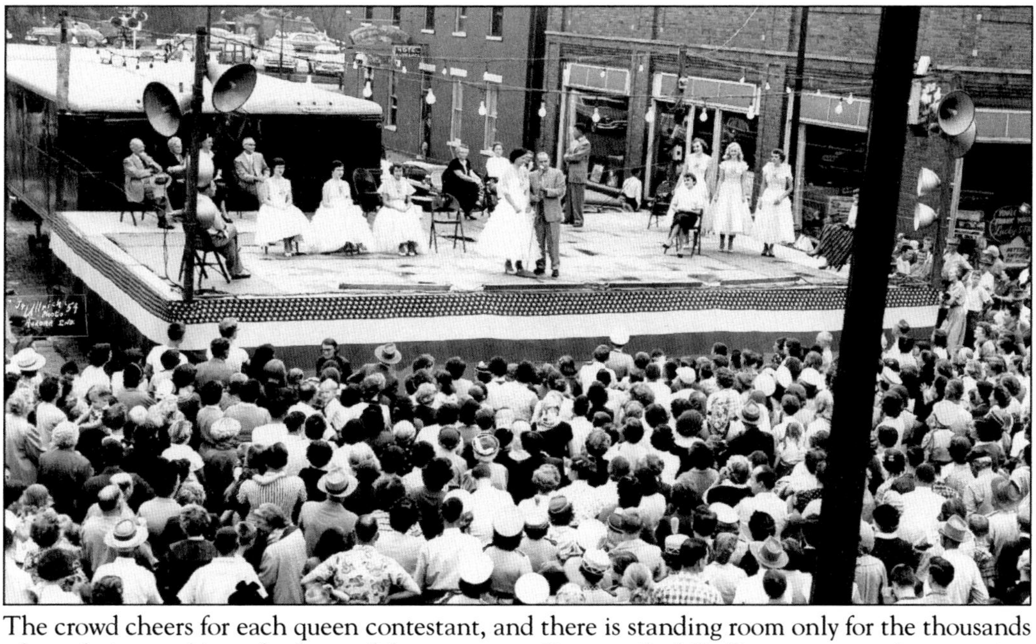

The crowd cheers for each queen contestant, and there is standing room only for the thousands of people jammed into the front of the stage area. Onstage Earl Huffman, master of ceremonies, questions a candidate. Judged on poise and response to questions about her plans for the future, Bonnie Craig, of Dillsboro, was crowned queen in 1954. (J. F. Ullrich collection.)

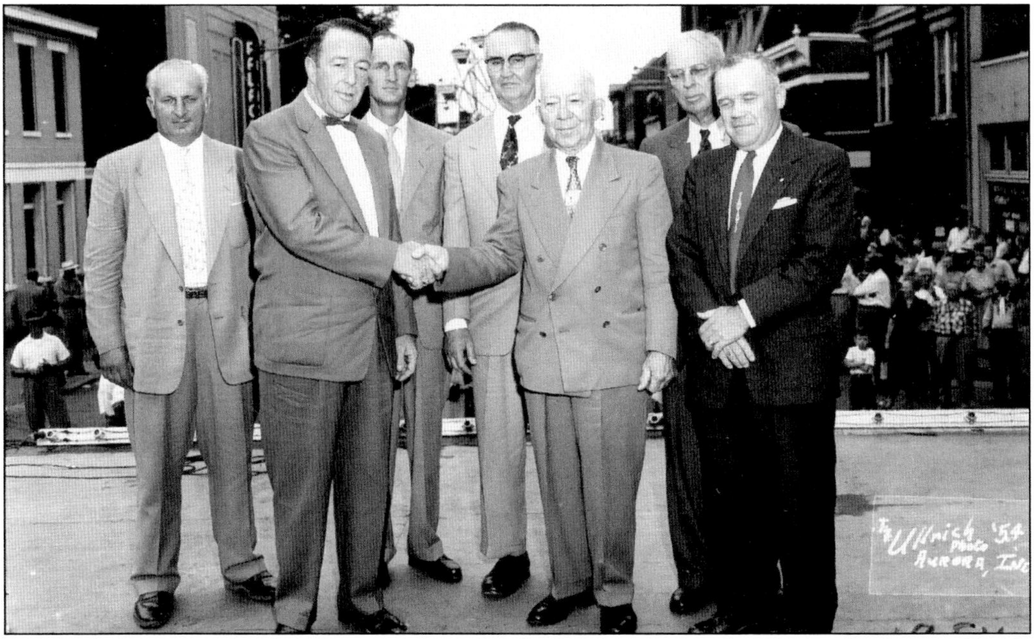

By October 1, 1954, Highway U.S. 50 was finally finished so local and state highway officials took part in dedication ceremonies Saturday morning before the parade. From left to right are H. W. Steigerwald, past chamber of commerce highway committee chairman; Virgil "Red" Smith, state highway commission; Carrol LaGrand, engineer; Harry Rullman, chamber of commerce; Aurora mayor Frank Green; R. V. Achatz, Southern Indiana Telephone Company; and Robert Harrell, district highway engineer. (J. F. Ullrich collection)

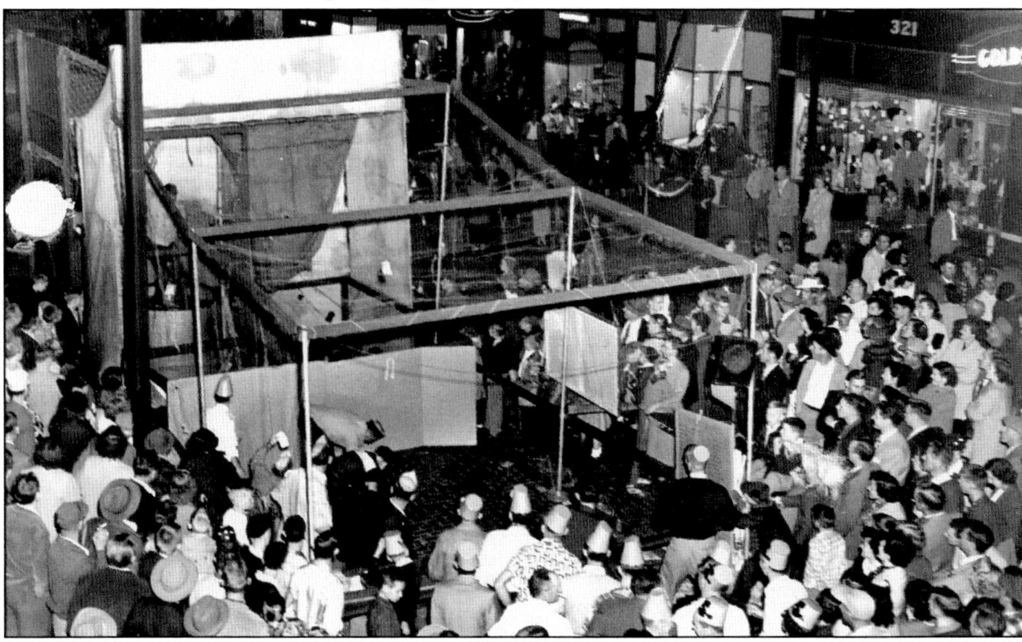

The Dunk-A-Lion project and wrestling matches were extremely successful fund-raisers for the Lions Club. The club donated money from one such event to help purchase life-saving equipment for the city. (J. F. Ullrich collection.)

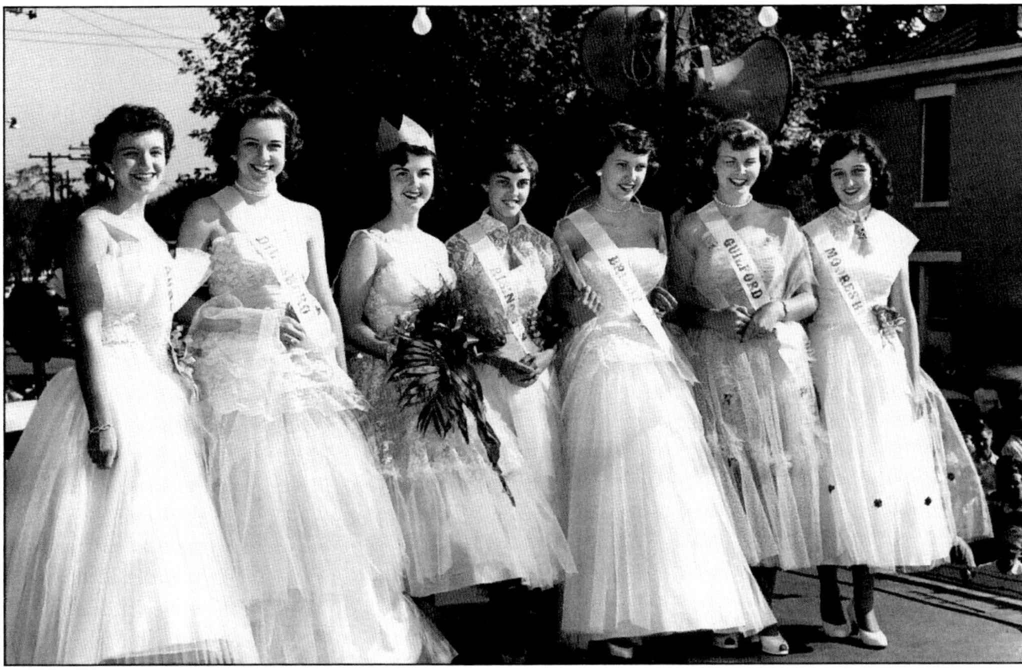

John Hicks, Rotary Club president, (not shown) placed the crown on 1955 fair queen, Janice Tomlin, a Lawrenceburg High School senior. Other candidates were Rosalie Scott, Dillsboro; Patti Page, Aurora; Verlene Holder, Rising Sun; Mary Cherry, Guilford; Jo Ann Holt, Moores Hill; and Maxine Smith, Bright. (J. F. Ullrich collection.)

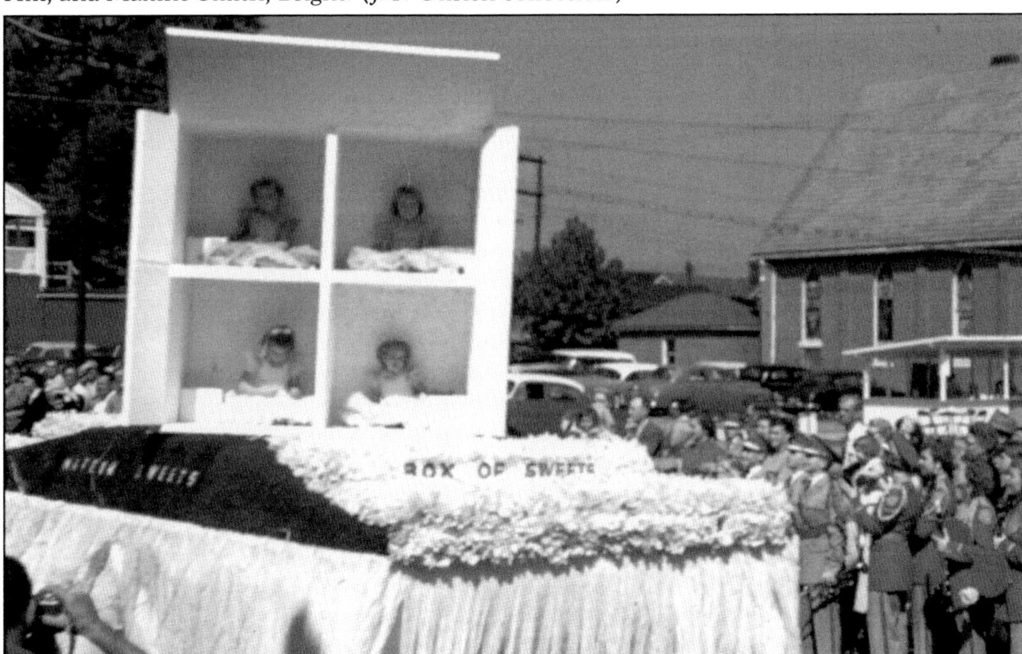

Young children pretend to be sweet pieces of candy in a box. In 1955, the National Container Corporation won the grand sweepstakes award of $50 for most outstanding float and was sponsored by Aurora Chamber of Commerce. (Tandy collection.)

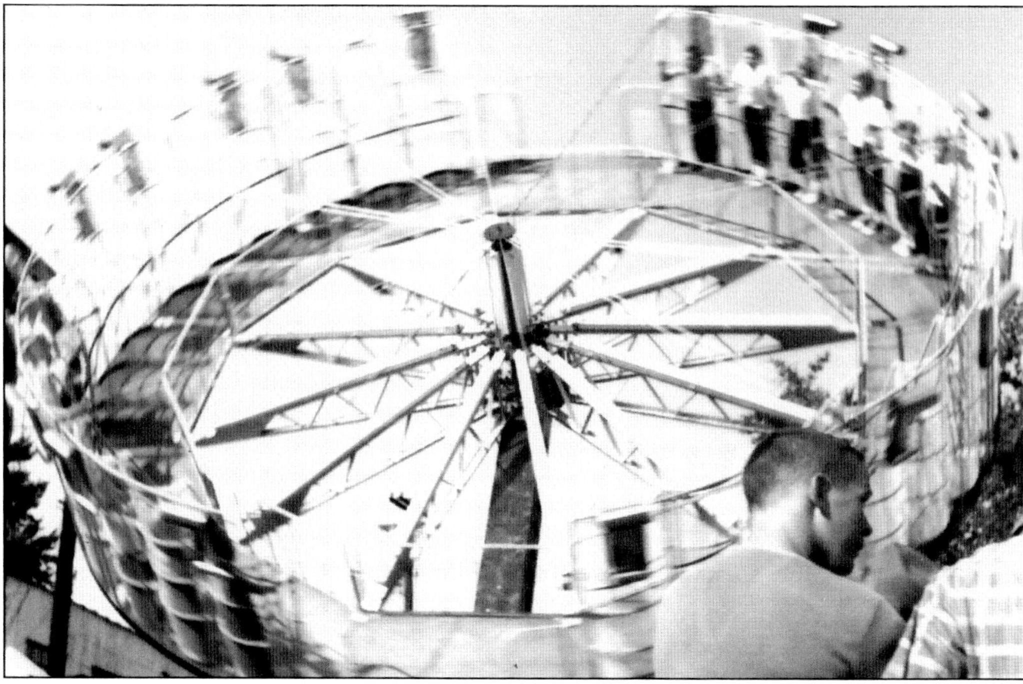

The Round-Up was a very popular ride for teenagers in 1956. Turning in a circle, the centrifugal force would hold the rider against the wall. It would then turn sideways, much to the delight of the fairgoers. (Weaver collection.)

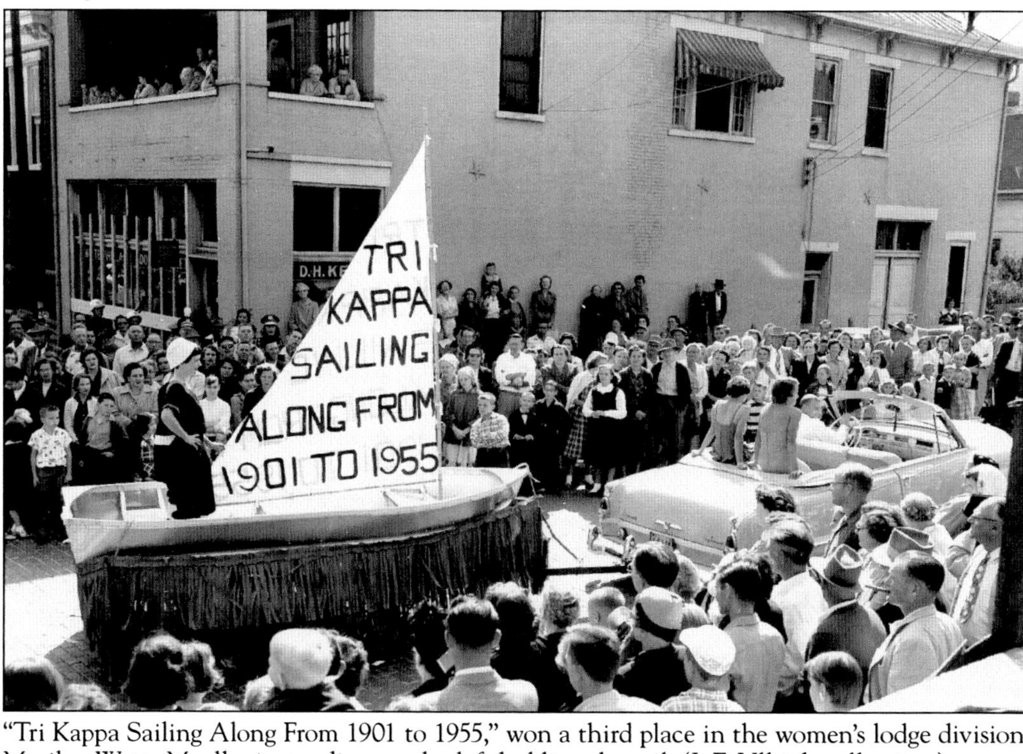

"Tri Kappa Sailing Along From 1901 to 1955," won a third place in the women's lodge division. Marilyn Watts-Moeller is standing on the left holding the sail. (J. F. Ullrich collection.)

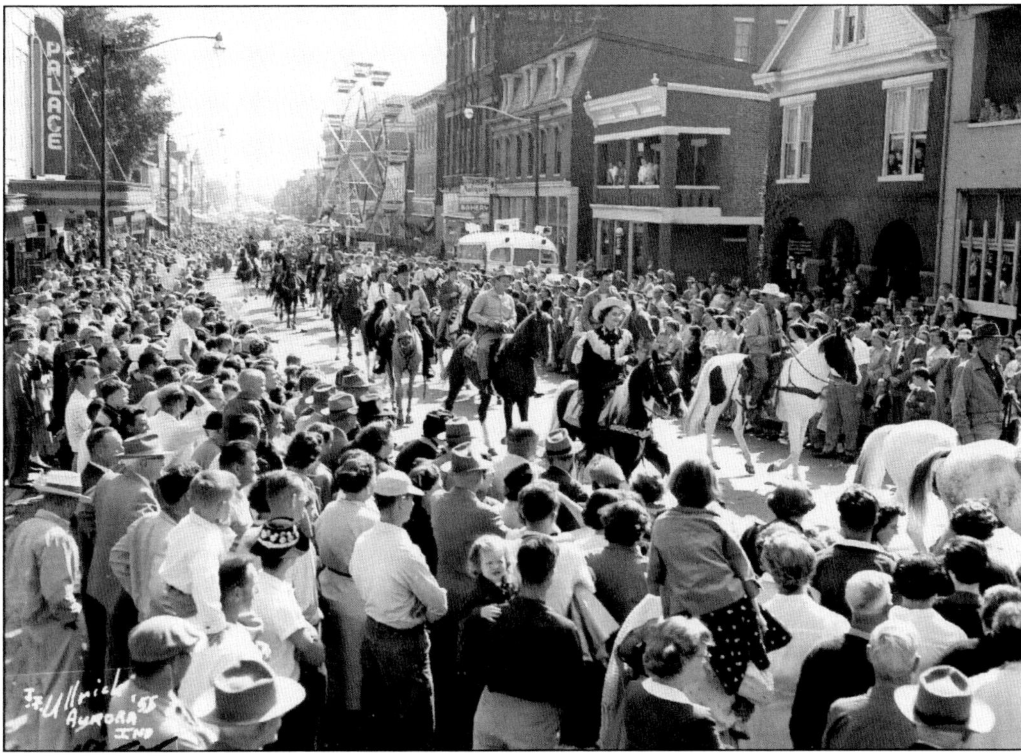

A large entry of mounted horses on parade make their way in front of the grandstand. (J. F. Ullrich collection.)

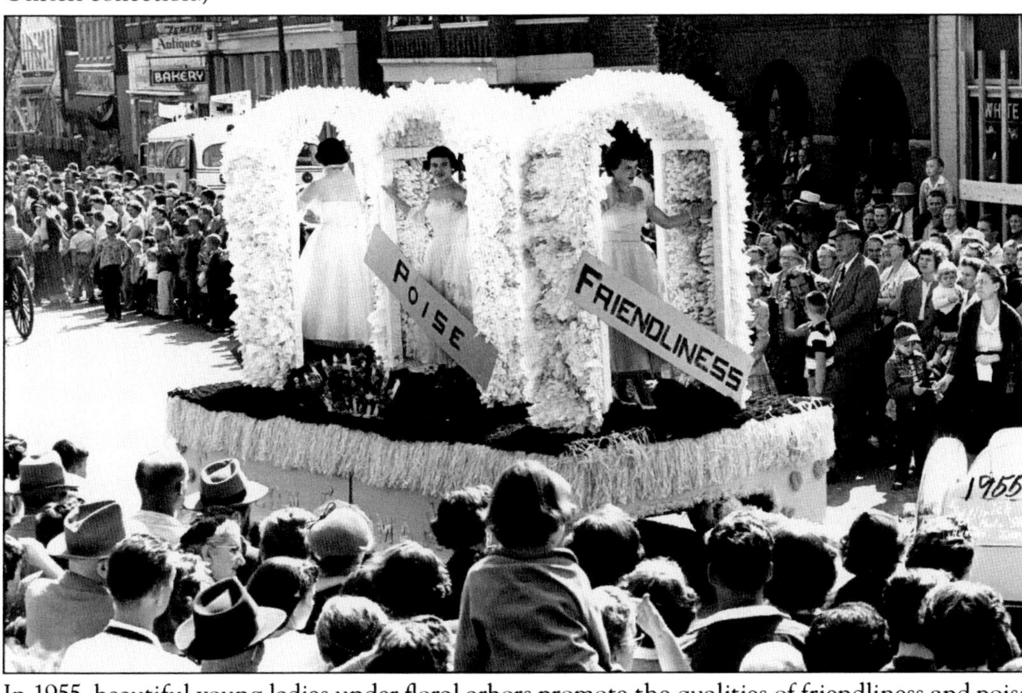

In 1955, beautiful young ladies under floral arbors promote the qualities of friendliness and poise as they pass in front of the stage. (J. F. Ullrich collection.)

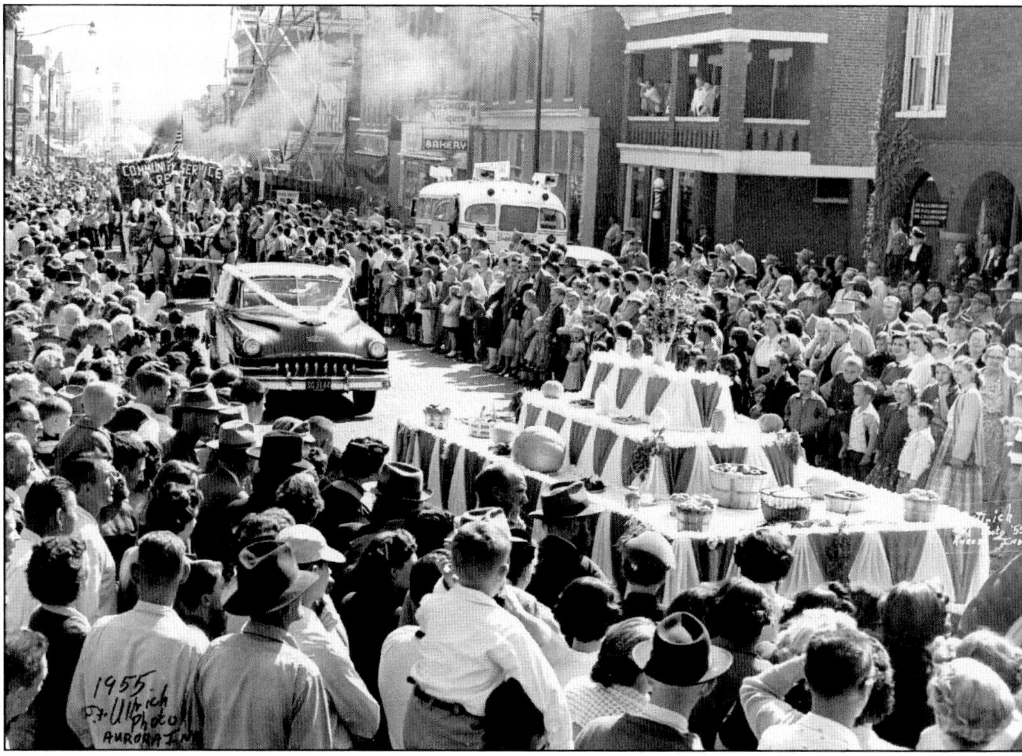

A float displaying various farm crops represents the true meaning of the farmers fair. (J. F. Ullrich collection.)

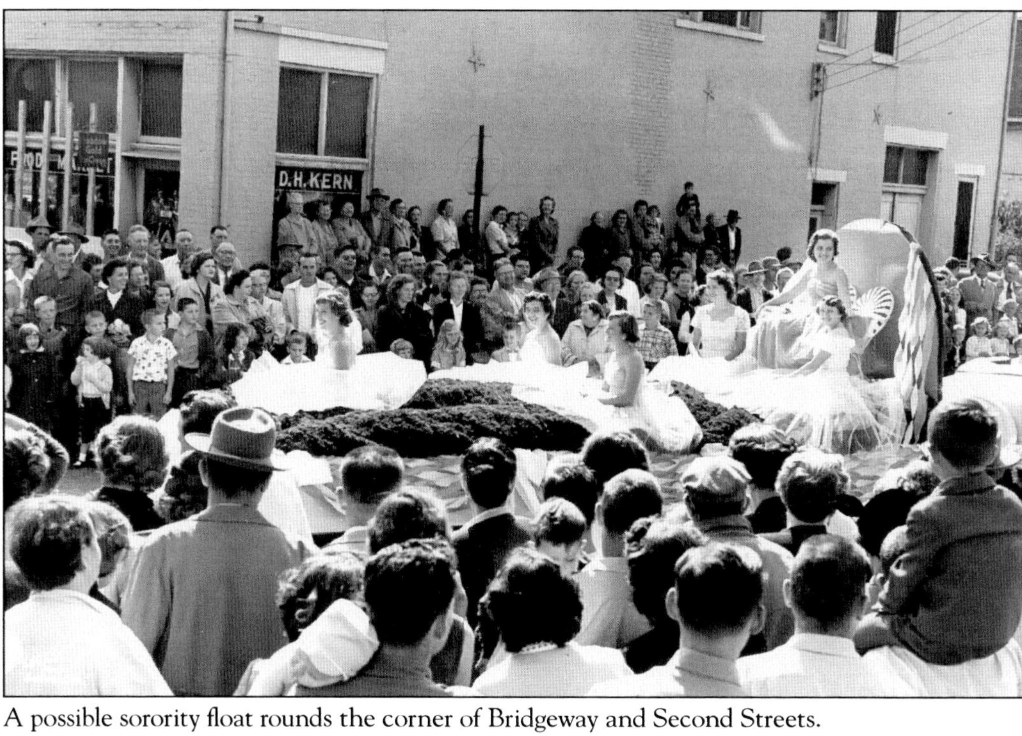

A possible sorority float rounds the corner of Bridgeway and Second Streets.

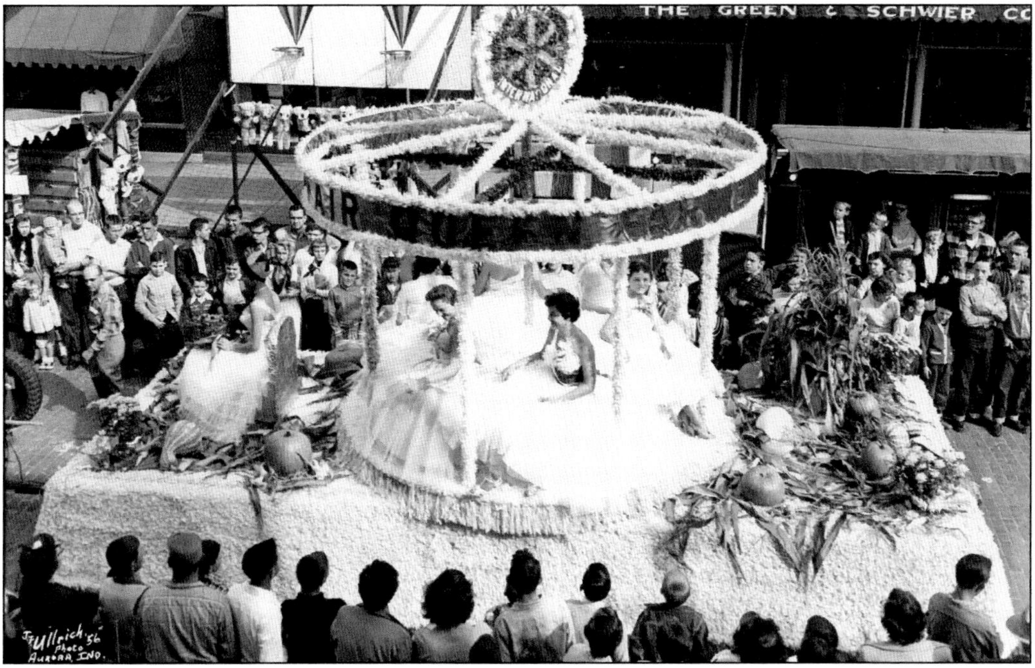

The 1956 queen's float, constructed by members of the Aurora Rotary Club, was unique. Queen Alyce (Nixon) Laker, Lawrenceburg, sat in front of a merry-go-round upon which the entire queen's court was seated. The carousal turned as strains of the calliope were heard. Candidates were Joy Beach, Aurora; Shirlee Sellers, Dillsboro; Dorothy Knue, Guilford; Patty Grace, Moores Hill; Forsteen Ratliff, Bright; and Rita Whitham, Rising Sun. (J. F. Ullrich collection.)

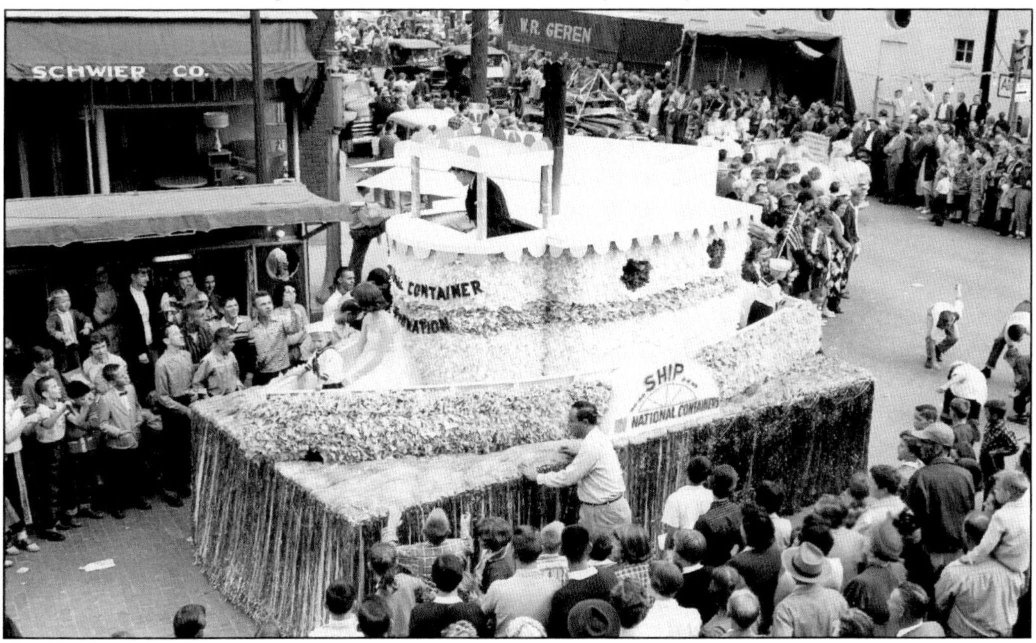

The National Container Corporation of Indiana won the first award in the industrial division of the parade. This entry depicted a river side-wheel steamboat, complete with beautiful girl passengers. Terry Trester was the pilot in the pilothouse. (J. F. Ullrich collection.)

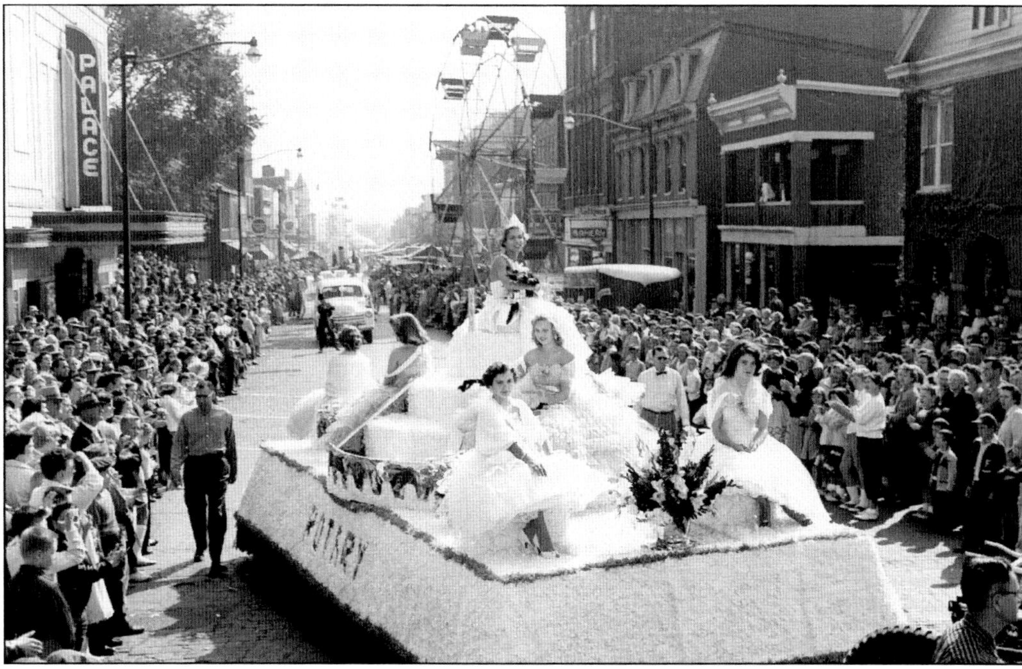

Queen Linda Schuler of Aurora is seated on the float's top tier. Fellow contestants were Susanna Terrill, Lawrenceburg; Mary Miller, Guilford; Glenda Schuler, Moores Hill; Wilma Washnock, Patriot; Betty Ballard, Dillsboro; Glenda Brook, Bright; and Anna Louise Worthington, Rising Sun. (J. F. Ullrich collection.)

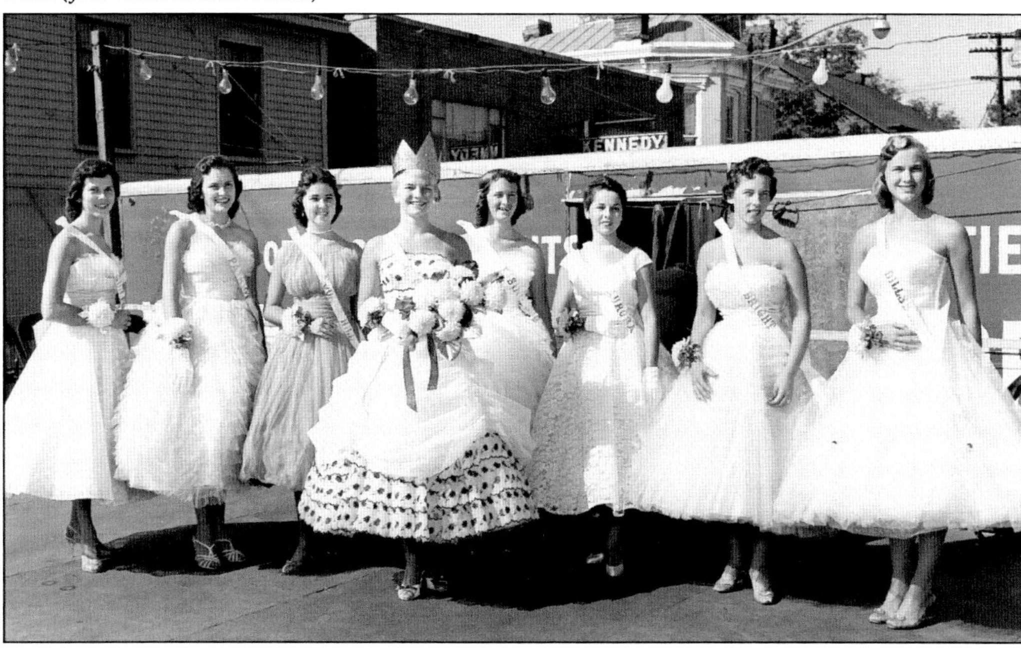

Celebrating the fair's 50th anniversary are the queen contestants Mary Knue, Guilford; Betty Martin, Patriot; Barbara Carter, Moores Hill; queen Mary Jane Terrill, Lawrenceburg; Ellen Hehe, Rising Sun; Patty Karsteter, Aurora; Jeanne Henderson, Bright; and Helena Cassidy, Dillsboro. (J. F. Ullrich collection.)

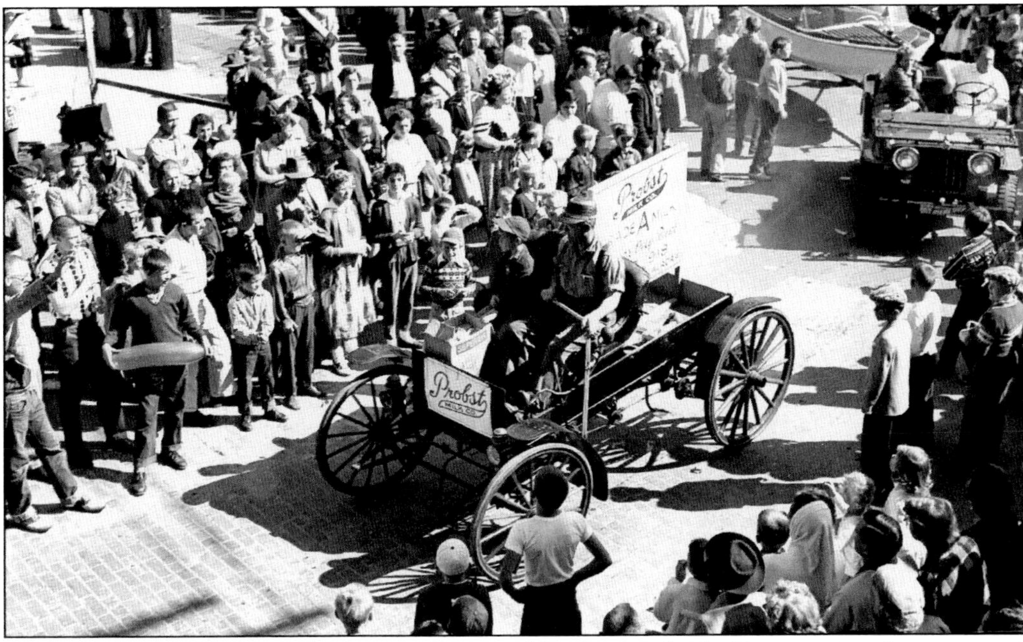

Edward Probst, with son Rick, drives a Sears and Roebuck vehicle. By removing the rear seat it could become a delivery wagon. There is a tiller for steering. The wooden-spoke wheels carry 38-by-2-inch solid rubber tires while the power comes from a small, 14-horsepower engine. The rear wheels were driven by a dual chain drive. This car is now on display at the state museum in Indianapolis. (J. F. Ullrich collection.)

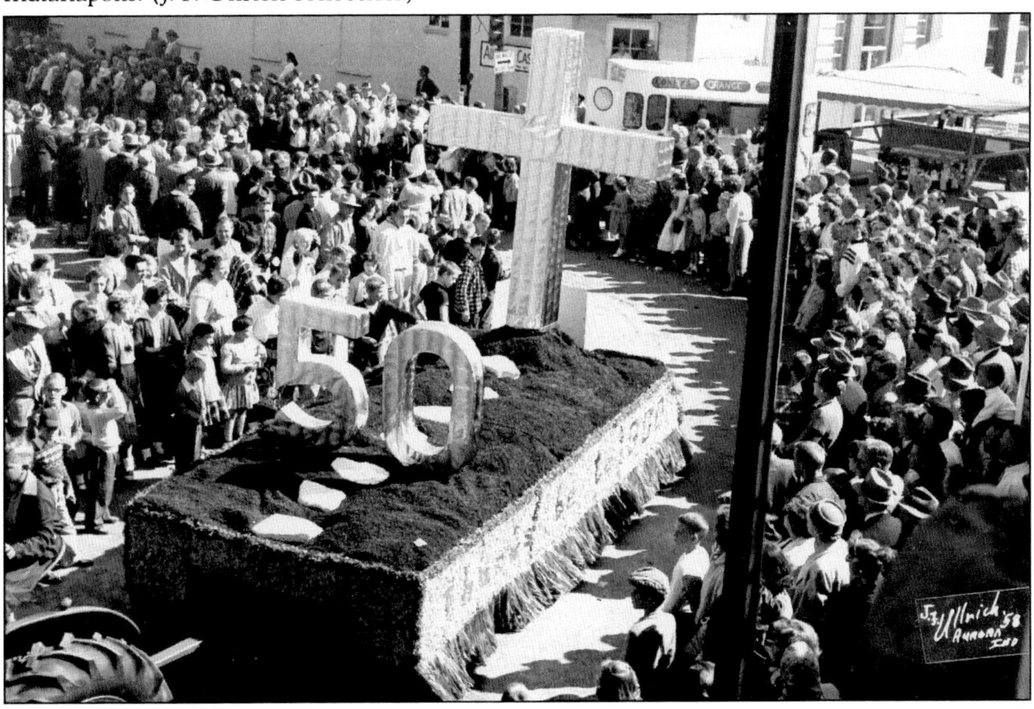

St. Johns Lutheran Church, Aurora, displayed a large cross on a float celebrating the farmers fair's 50th anniversary. The sides of the float say, "Thanks Be To God." (J. F. Ullrich collection.)

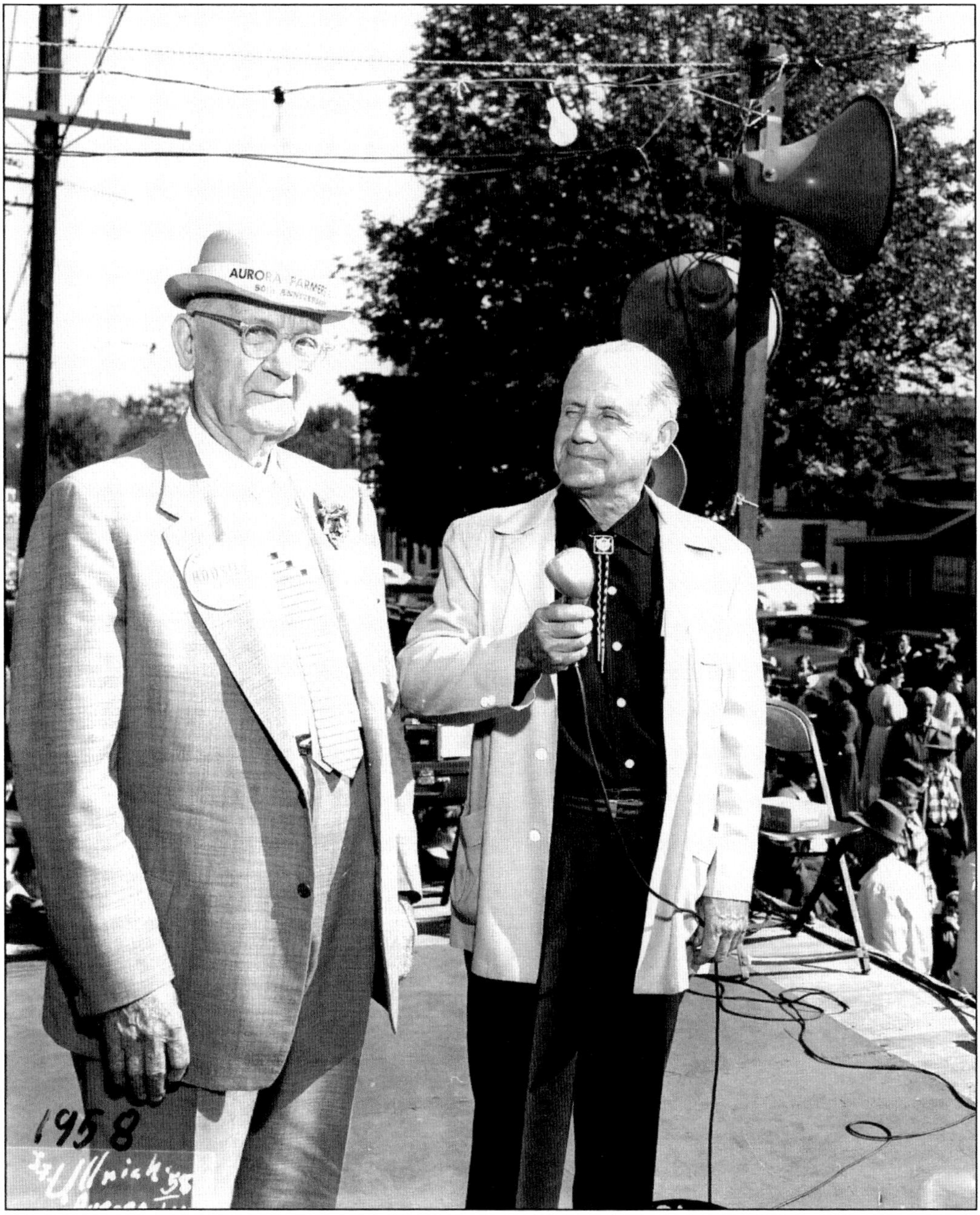

Master of ceremonies Earl Huffman interviews Robert L. Johnston. Johnston's work in the fair was so prominent that in 1916 he was elected president of the farmer fair board and retained that office for well over 40 years. He was recognized as a wonderful citizen and businessman and wanted to give the thousands who visited Aurora during the fair a program of entertainment. Johnston believed and often said, "What is more valuable to any community than to have friends, that remember and love the town of their childhood." (J. F. Ullrich collection.)

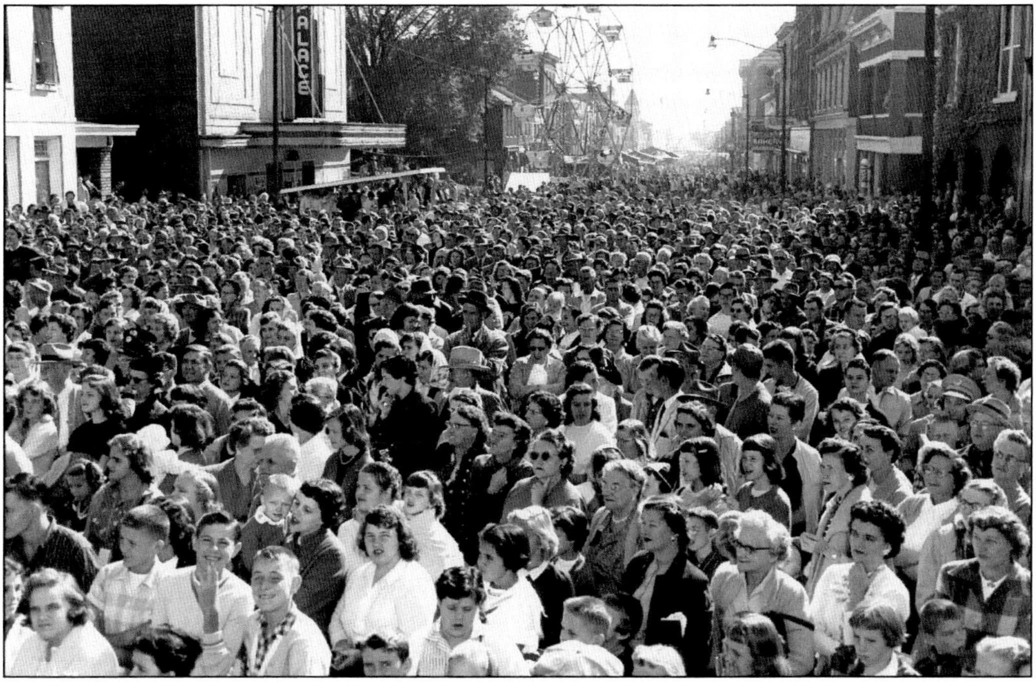

A crowd, as far as the eye can see down Second Street, stands in front of the grandstand watching the free entertainment provided. (J. F. Ullrich collection.)

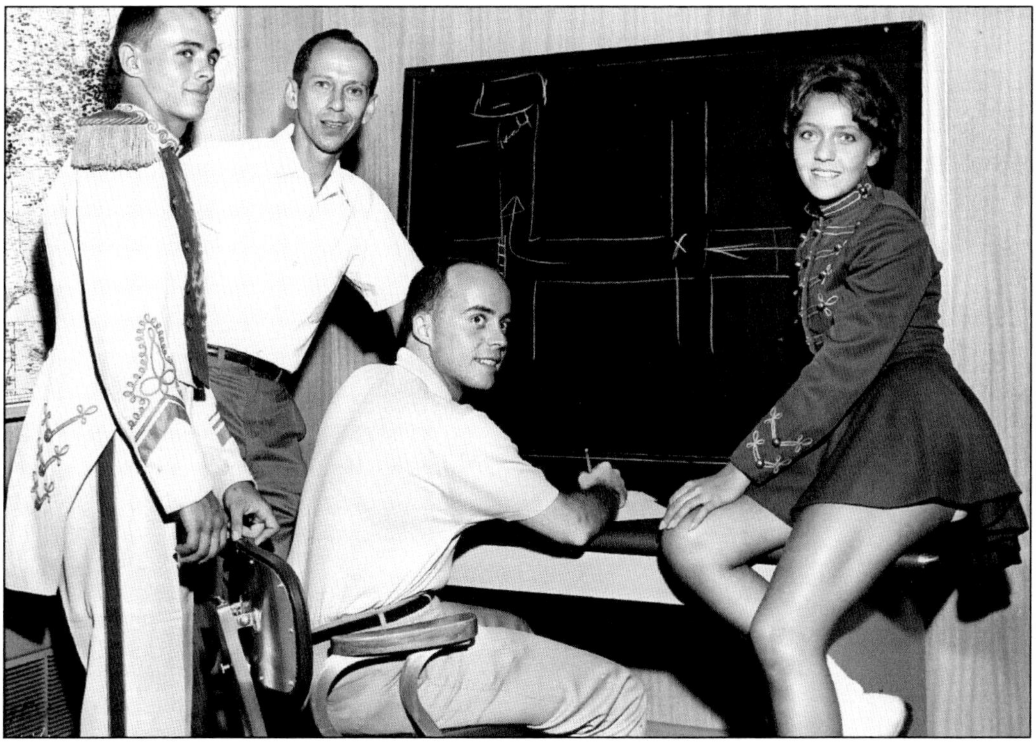

Mapping out the parade route for the band, from left to right, are drum major Jay Wright, band director Bill Platt, Jim Dartnell, and majorette Sonya Schuler. (J. F. Ullrich collection.)

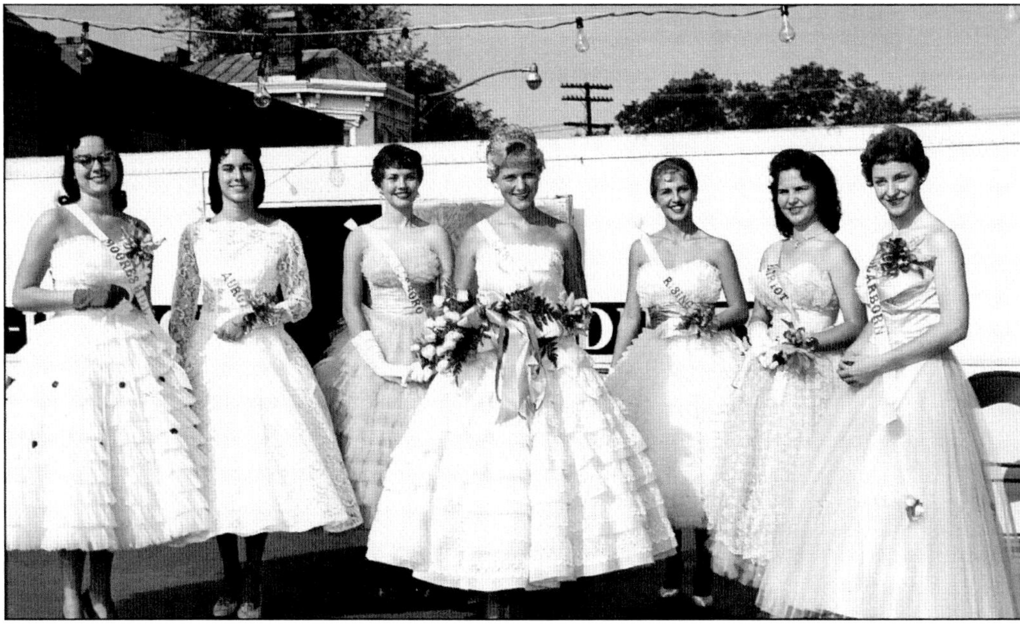

On October 3, 1959, Sue (Haag) Karsteter, of Lawrenceburg, is crowned queen. The queen's court, from left to right, includes Janice Bolin, Moores Hill; Pam Porter, Aurora; Retta Thayer, Dillsboro; runner-up Babetta Huffman, Rising Sun; Shirley Gould, Patriot; and Irene Spaulding, North Dearborn. This was North Dearborn's first year due to the consolidation of the Guilford and Bright high schools. (J. F. Ullrich collection.)

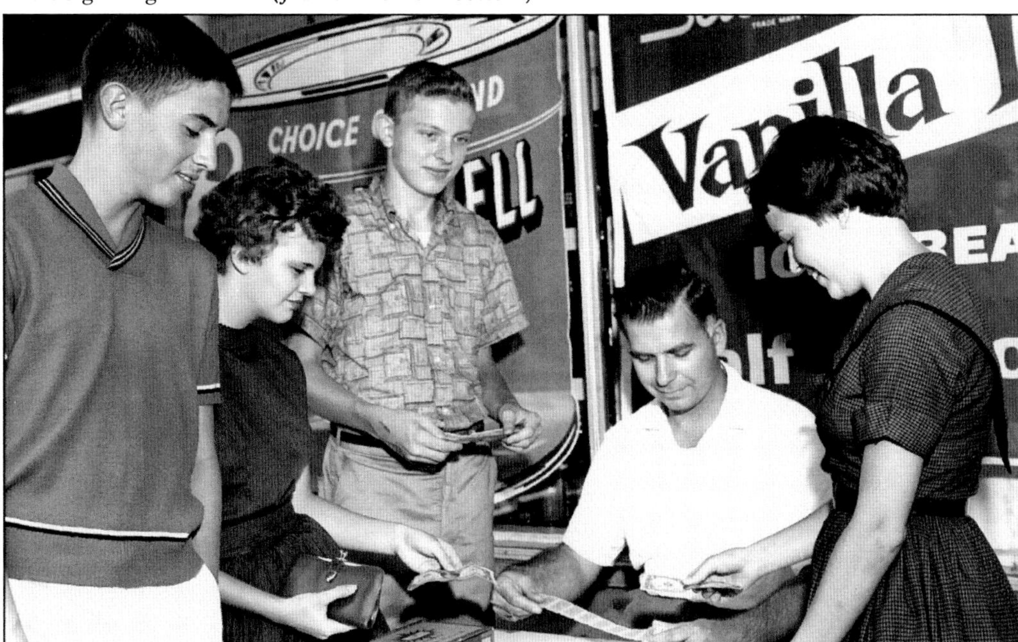

Seated above, John Wunderlich, a charter member of the Aurora Lions Club, is shown selling advanced special-price ride tickets for the dates September 29 and 30, 1959. Purchasing tickets are, from left to right, Ronny Kemper, Jane Cheek, Marty Rahe, and Faith Kemper. Advanced ride tickets were sold at half price, or eight tickets for $1. (J. F. Ullrich collection.)

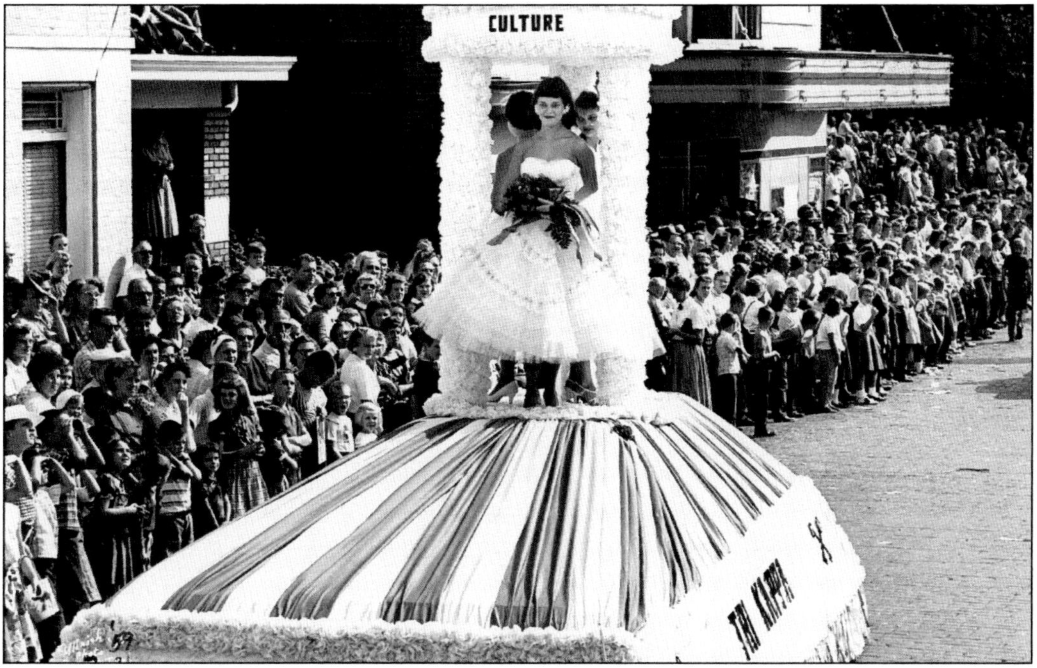

On October 3, 1959, the first award in the women's organization class went to this Tri Kappa float. The beautiful girls represented the goals of the sorority: charity, cultures, and education. The girls were Sharon Bradford, Jane Linkmeyer, and Jane Cheek. The Tri Kappa sorority was always in heavy competition with the Phi Beta Psi sorority for the best float display. (J. F. Ullrich collection.)

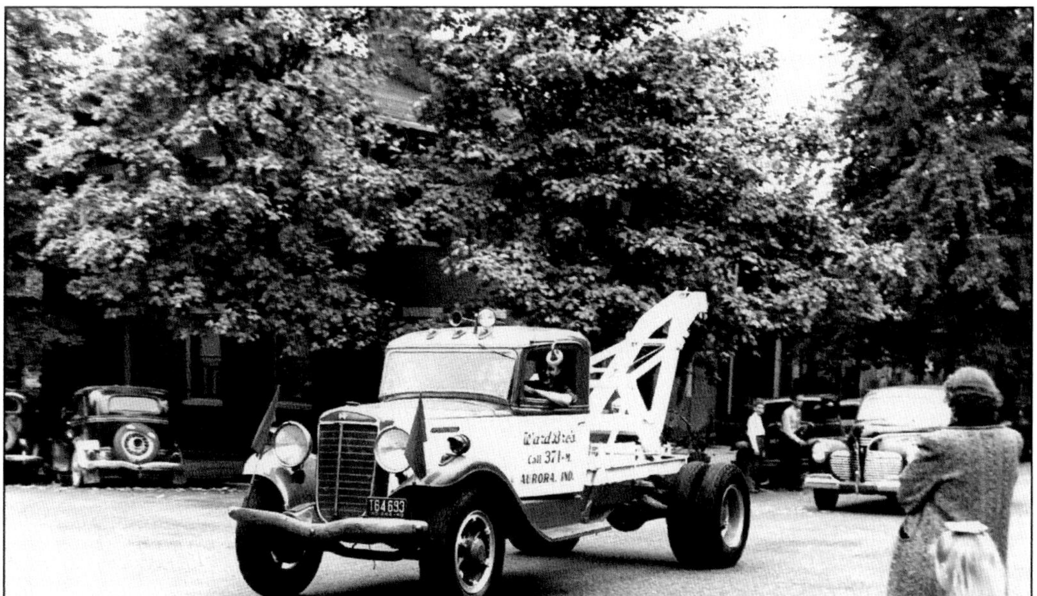

This 1928 International Model C-40 tow truck belonged to Ward Brothers Garage and is driven by employee L. B. "Bo" Johnston. Bo liked to bring his young son Charles with him to car accidents to help sweep up the glass and debris. (C. Johnston collection.)

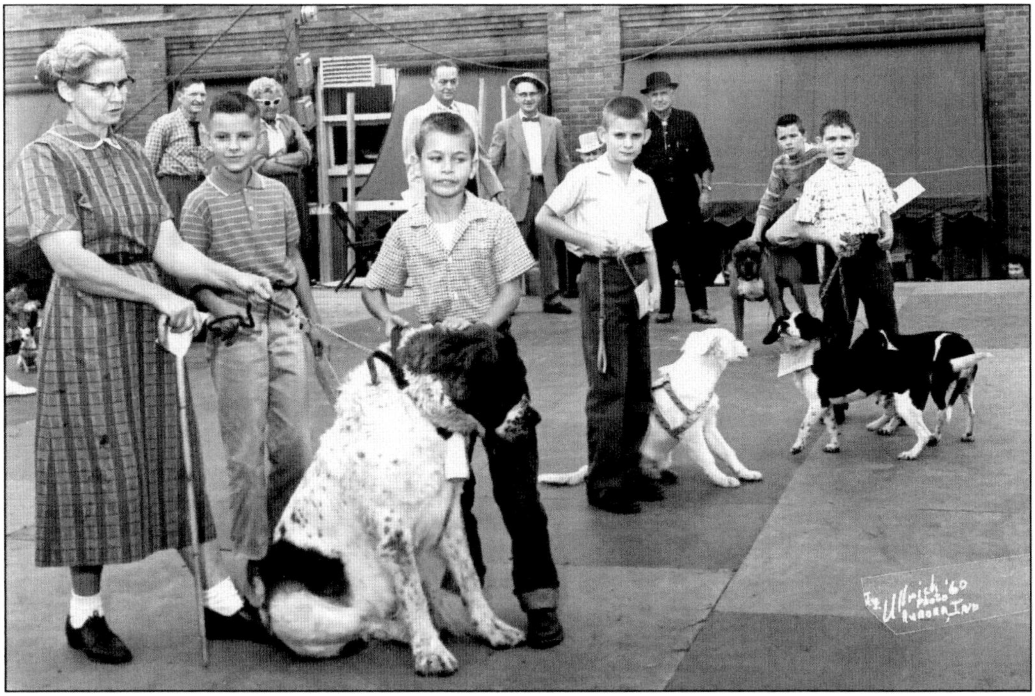

Helen Probst, Ron Johnston, and Rick Probst try to manage a St. Bernard in the biggest dog contest. John Uhlmansiek, Ron Draper, and Terry Klueber are the other contestants onstage. (J. F. Ullrich collection.)

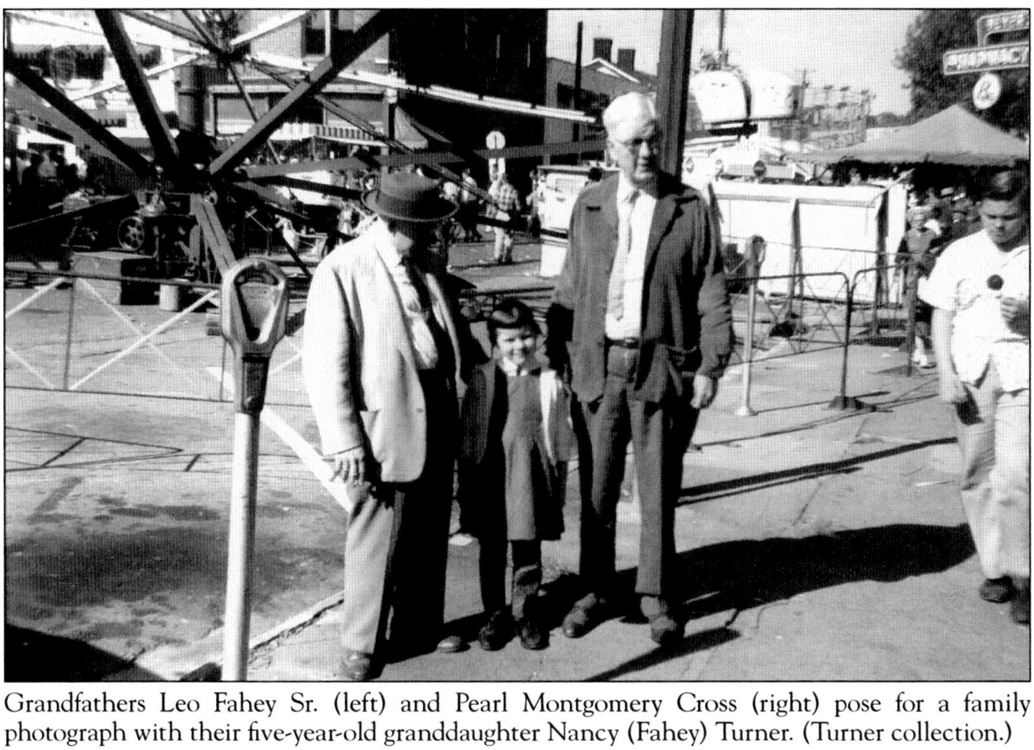

Grandfathers Leo Fahey Sr. (left) and Pearl Montgomery Cross (right) pose for a family photograph with their five-year-old granddaughter Nancy (Fahey) Turner. (Turner collection.)

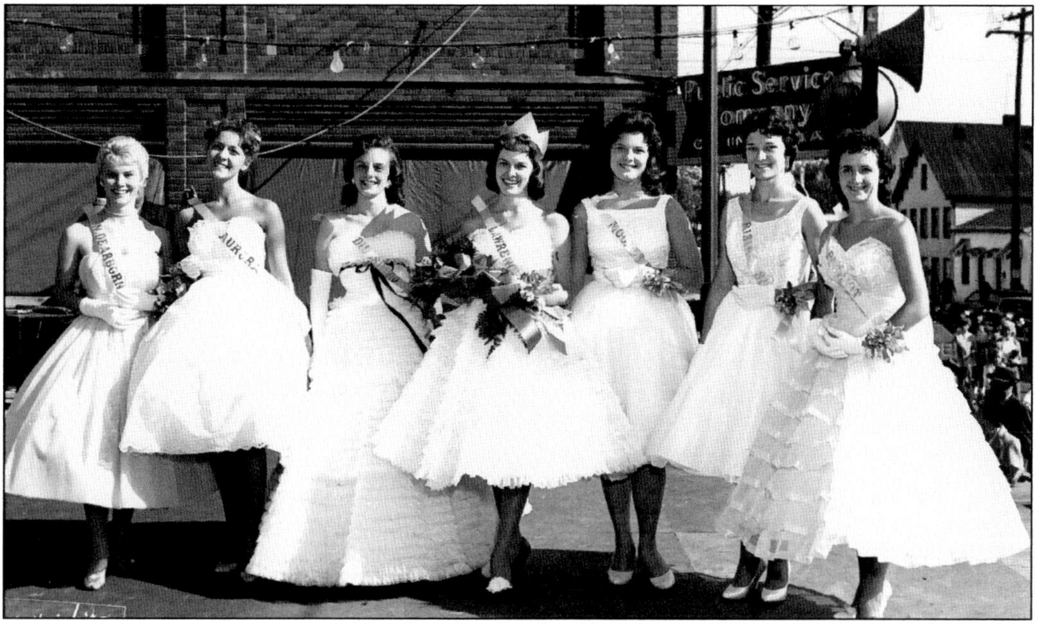

From left to right, the queen contestants are Judy Downey, North Dearborn; Sonya Schuler, Aurora; Barbara Ashcraft, Dillsboro; queen Cecilea Estes, Lawrenceburg: Ruth Beard, Moores Hill; Marcia Kirkpatrick, Rising Sun; and Saundra Graves, Patroit. (J. F. Ullrich collection.)

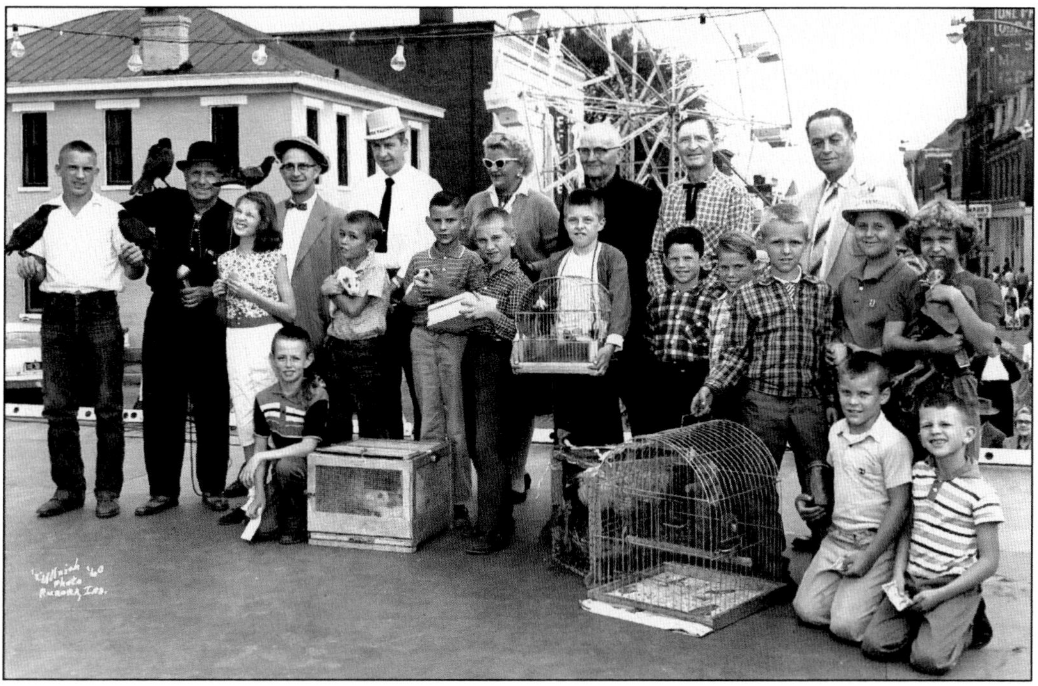

Winning the special prize award with his three pet crows is Barry Kimble of Holton. Also judged were Theresa Cato and her pigeon Frankie. Oscar the guinea pig is held by Rick Probst of Aurora. Other winners were Ronald Johnston, Aurora, with his white guinea pig; Mike Peters, Aurora, with Polly Parrott; and Gregory Riggs, Aurora, with Hercules the opossum. (J. F. Ullrich collection.)

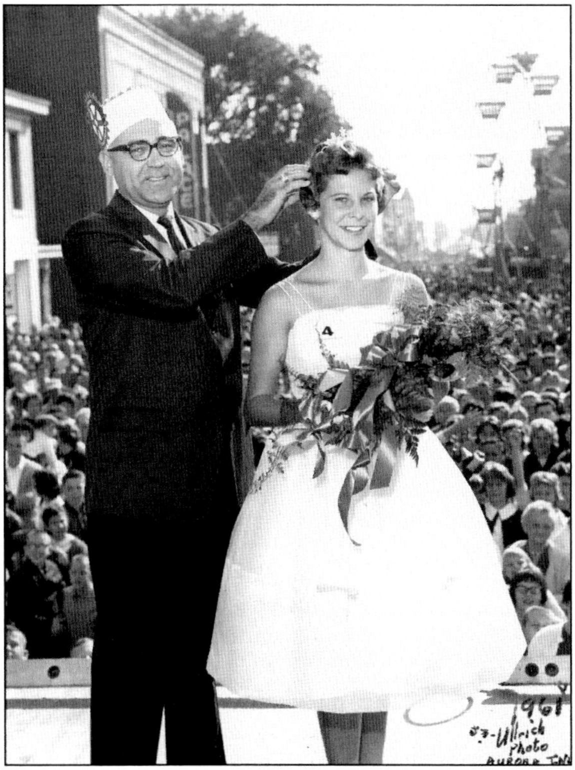

Dianna Bondurant, 17, a senior from Rising Sun High School, was crowned queen of the Aurora Farmers Fair by Don Hammer, president of the Rotary Club, before the parade on October 7, 1961. (J. F. Ullrich collection.)

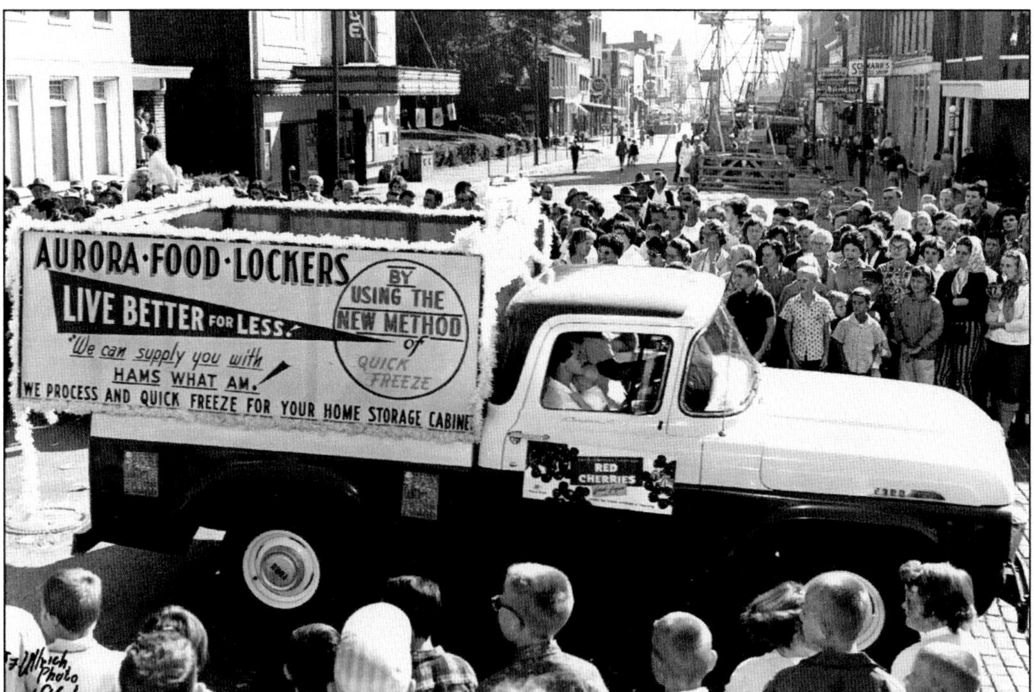

The Aurora Food Locker company truck owned by Robert L. Johnston makes its way along the parade route. Chris Thieman was a store manager. (J. F. Ullrich collection.)

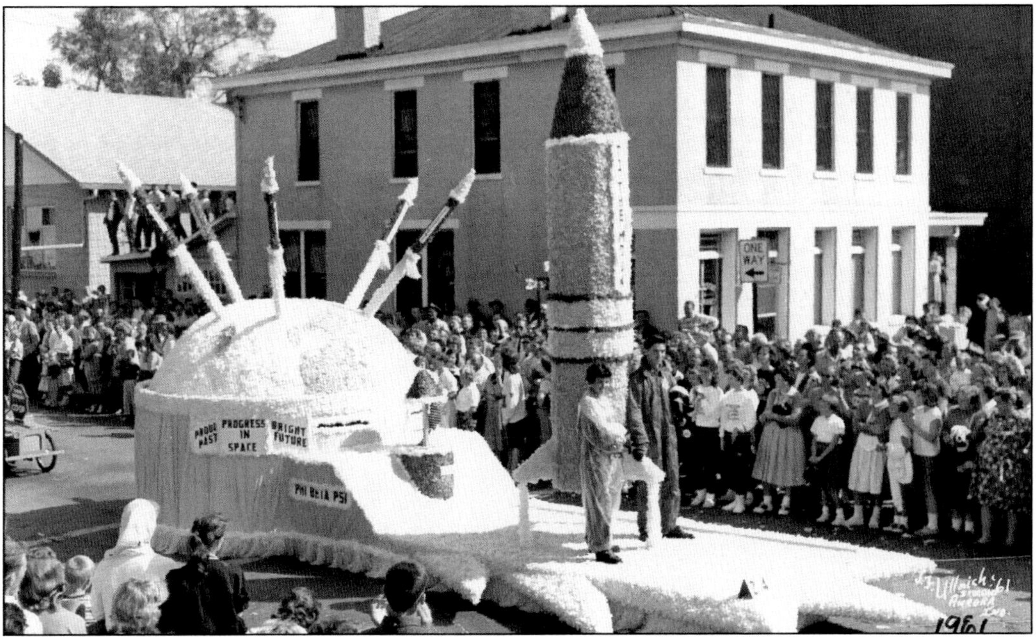

Progress in Space was the theme of the 1961 fair. The Phi Beta Psi sorority received the sweepstakes prize for best float. Standing on the float are Ron and Faith Kemper. The sorority received $50 and a trophy. (J.F. Ullrich collection.)

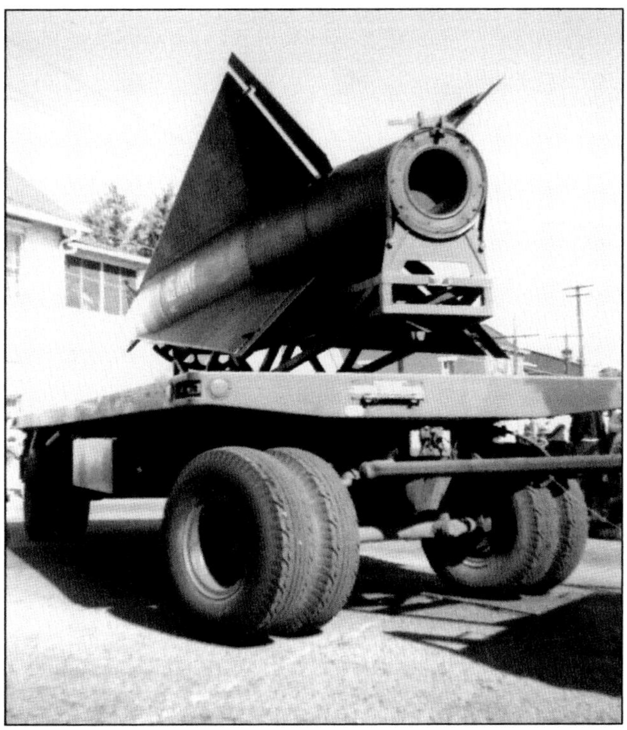

Keeping with the theme Progress in Space, a Nike Hercules missile display and a color guard from the Dillsboro Nike base led the parade. Nike was the name given to the world's first successful, widely deployed, guided surface-to-air missile system. The Hercules missile saw service starting in 1958. It had a range of 75 miles and could carry nuclear or conventional explosive warheads. (Ravenscraft collection.)

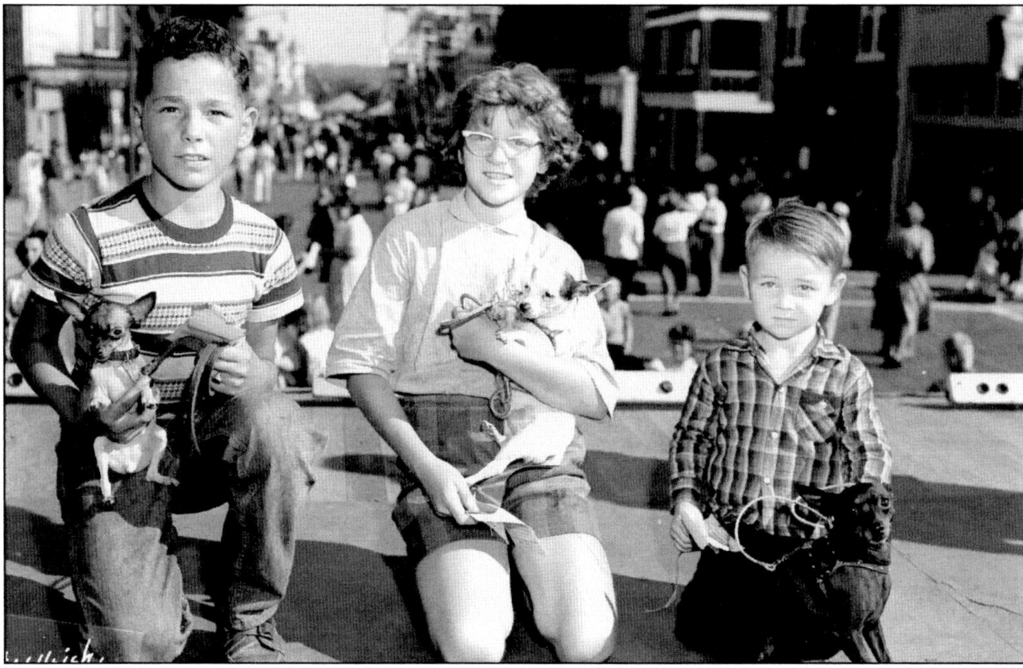

Winners of the smallest dog contest were Donnie Earls, Debbie Knippenberg, and Kevin Beard. (J. F. Ullrich collection.)

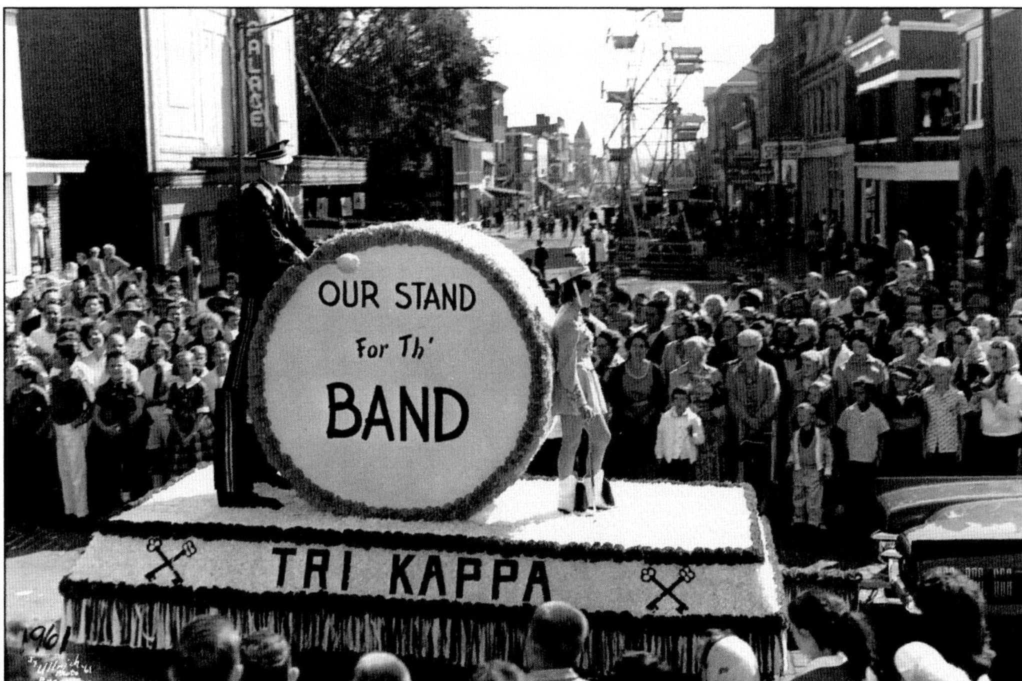

The Tri Kappa sorority constructed this float to help raise money for new band uniforms. Aurora has always had reason to be proud of its high school band. The school board was not allowed to provide either uniforms or transportation for the band, leaving it up to the parents association and other organizations to help raise funds. (J. F. Ullrich collection.)

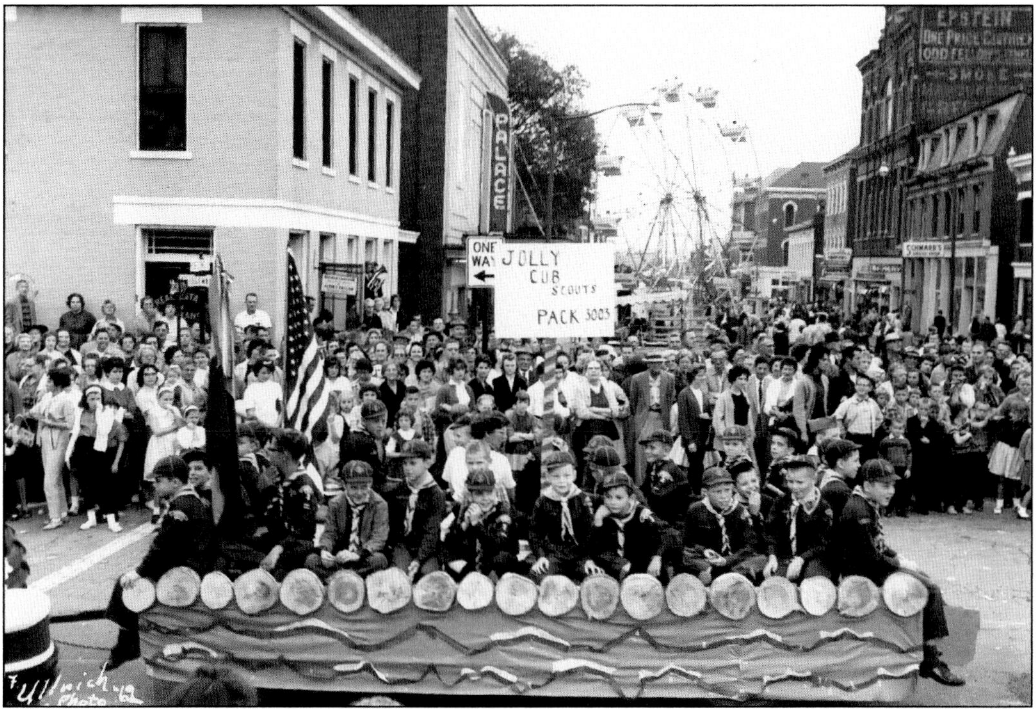

The Jolly Cub Scouts pass the stage on their display. (J. F. Ullrich collection.)

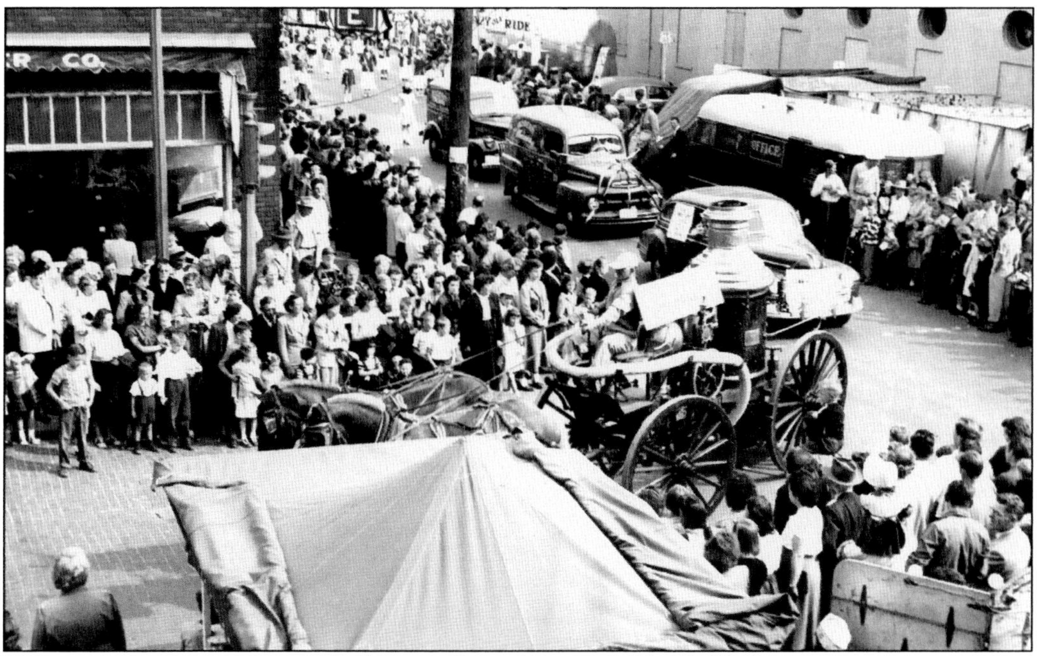

Horses pull the Aurora Fire Company No. 2 steam fire engine, built in 1895, for one last run down Second Street. The old engine was donated by Aurora on April 4, 1962, to the Cincinnati Fire Department Museum. Named the *Thomas Gaff*, in honor of the pioneer and Aurora distiller who donated the money for its purchase, it still stands on display at the museum. (S. Ullrich collection.)

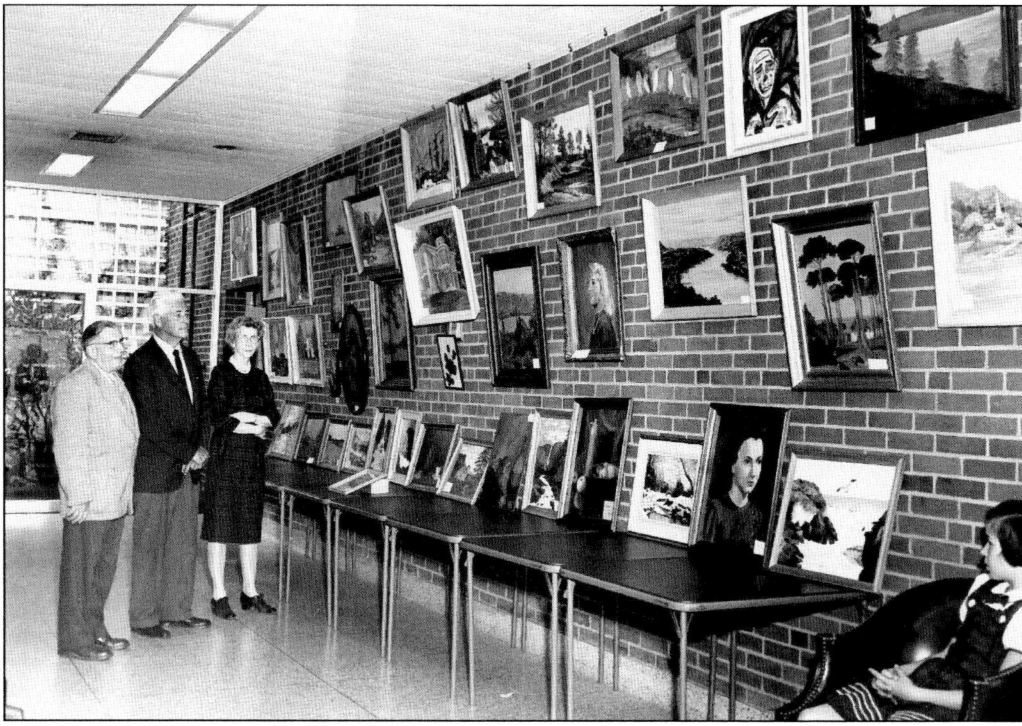

The First National Bank held an art exhibit during the 1962 fair. Charles Klingelhoffer, Bill Backman I, and Ester Roache are viewing the art exhibit. "Everyone was surprised that we had so many artists," said bank employee Ester Roache. (J. F. Ullrich collection.)

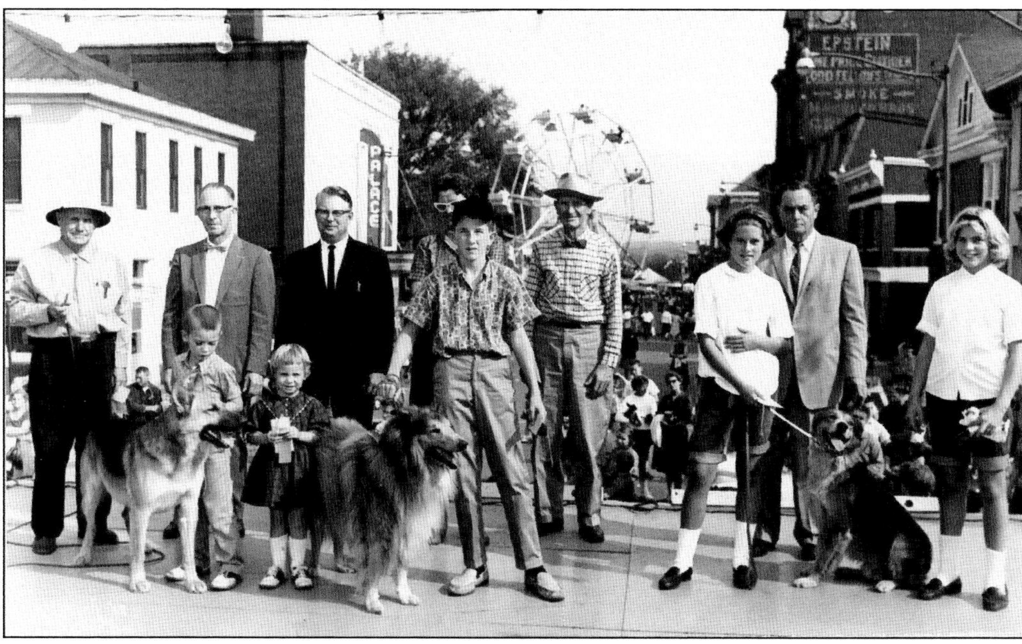

The large dog division winners were first prize, Terry and Tony Johnson with Shadow; second prize, Melvin Green with Prince; and third prize, Connie Klingelhoffer and Jean Barrott with Prince. (J. F. Ullrich collection.)

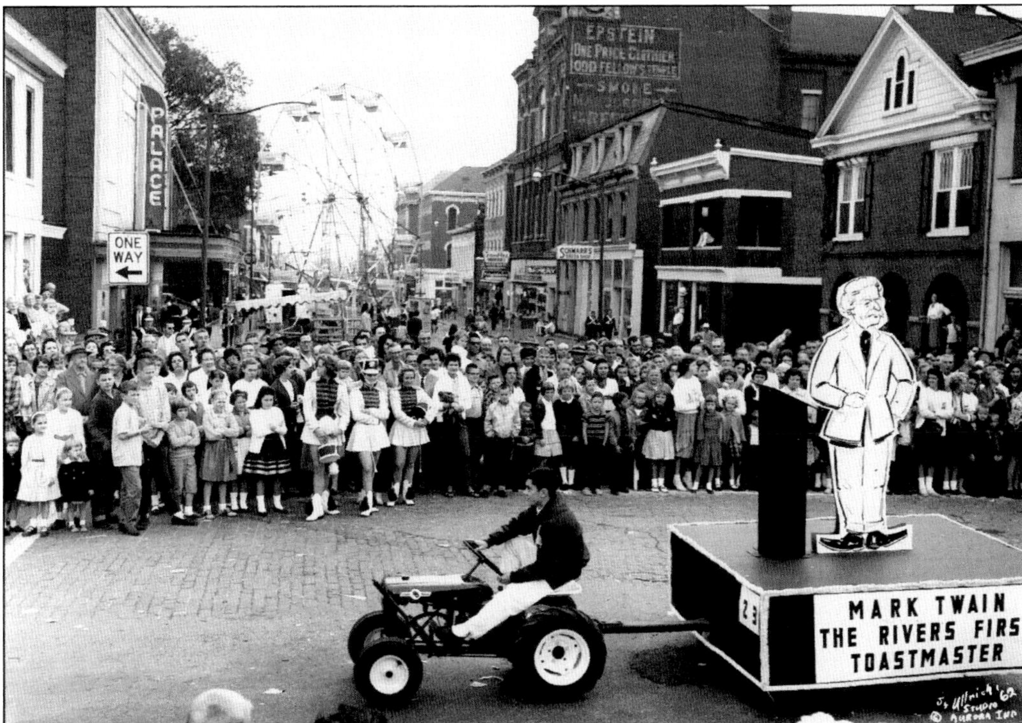

John Schuler pulls a small Mark Twain float up Bridgeway Street in front of the stage. (J. F. Ullrich collection.)

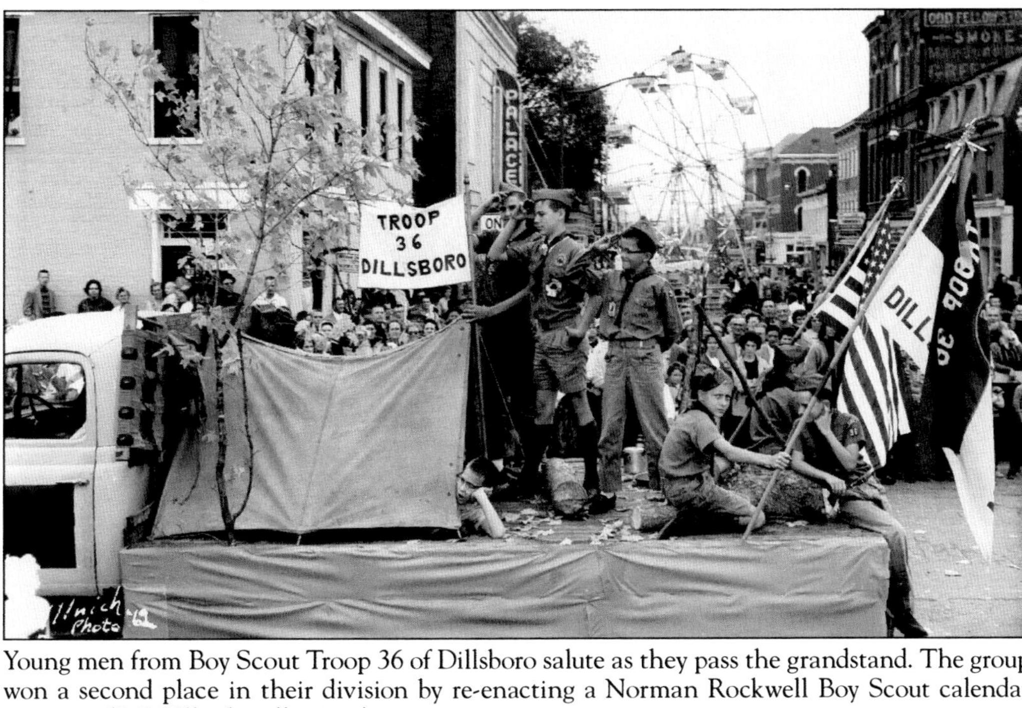

Young men from Boy Scout Troop 36 of Dillsboro salute as they pass the grandstand. The group won a second place in their division by re-enacting a Norman Rockwell Boy Scout calendar painting. (J. F. Ullrich collection.)

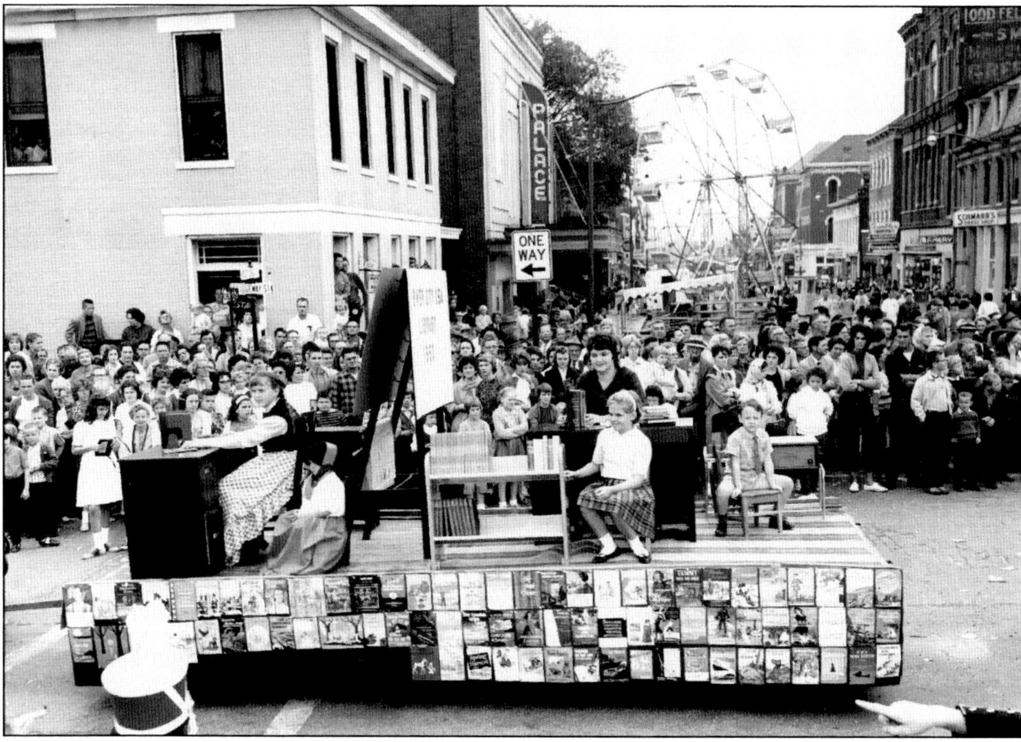

Aurora Library carries out the 1962 parade theme Yesteryear and Today. (J. F. Ullrich collection.)

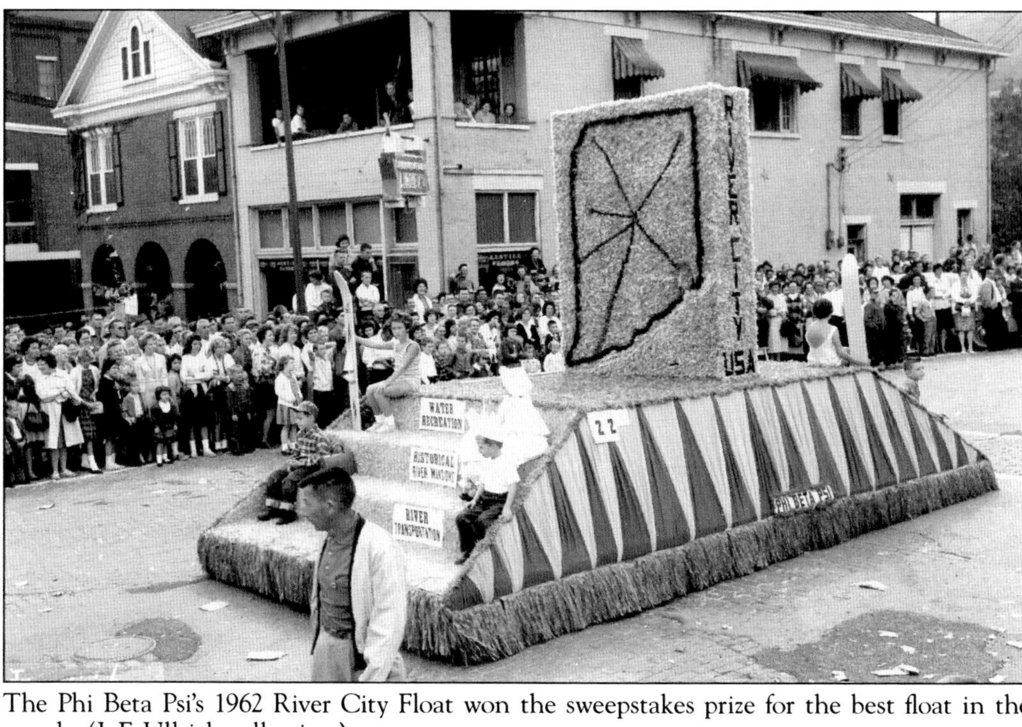

The Phi Beta Psi's 1962 River City Float won the sweepstakes prize for the best float in the parade. (J. F. Ullrich collection.)

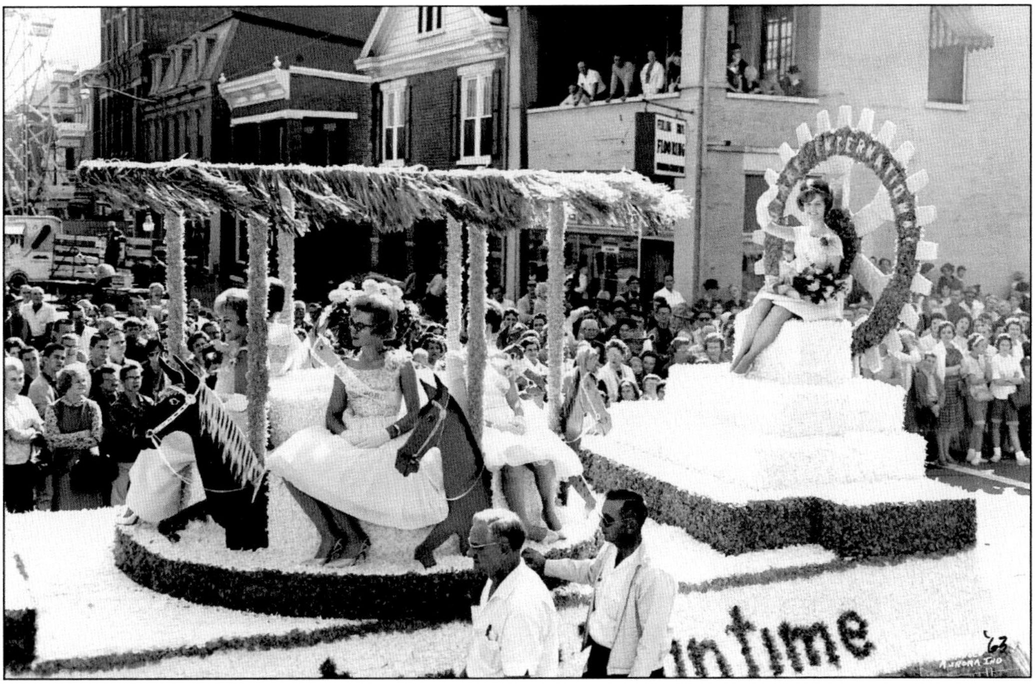

The candidates for the 1963 contest were queen Cheryl Barker, North Dearborn; Neysa Kay Ebel, Aurora; Carolyn Martin, Dillsboro; Adele Nickell, Lawrenceburg; Myrtle Robinson, Rising Sun; and Vera West, Moores Hill. (J. F. Ullrich collection.)

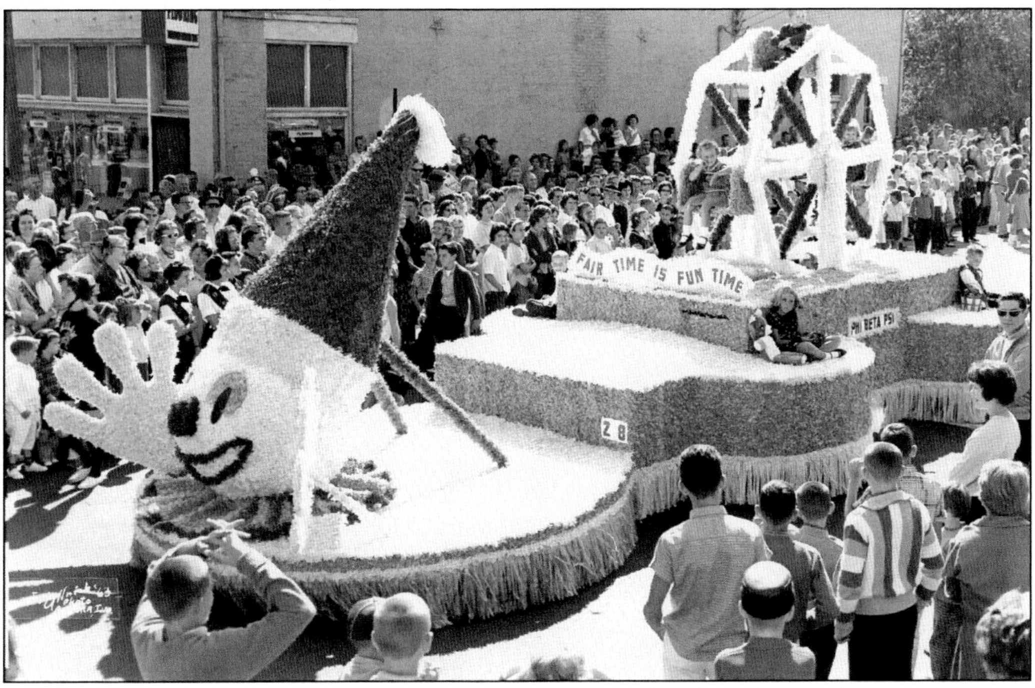

No doubt a prize winner, small children sit patiently throughout the parade on a make-believe Ferris wheel. Phi Beta Psi built this elaborate float for the 1963 parade. (J. F. Ullrich collection.)

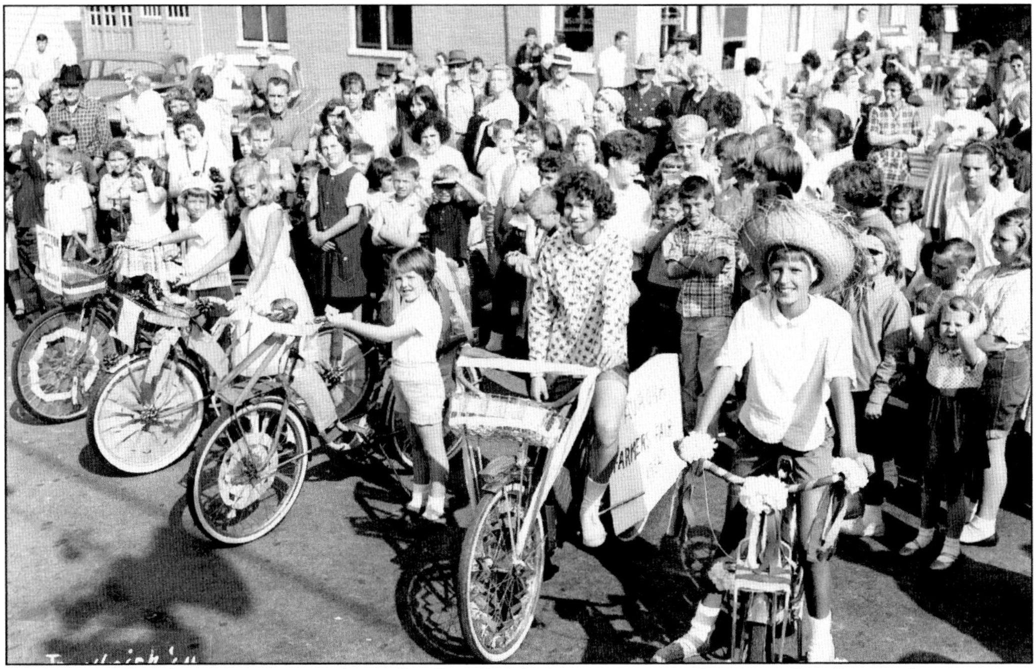

Sponsored by the First National Bank, the children were not only invited to bring their pets to be judged but also their bicycles. In the best decorated bicycle and rider class, from left to right, are Marilyn Sutton, Debora Conger, Susan Wunderlich, Lori Jo Skidmore, and Vickie Beam. (J. F. Ullrich collection.)

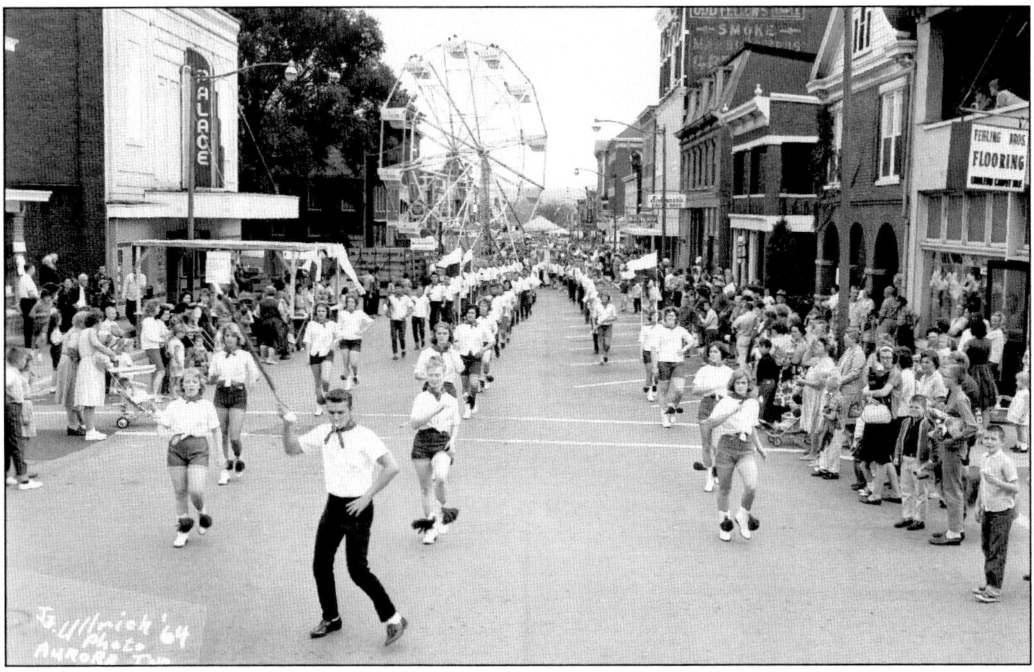

The head drum major, Mike Meyer, dances as he leads the Aurora band and the Friday afternoon pet parade up Second Street. Children with their pets always follow the band to the judging area at Second and Bridgeway Streets. (J. F. Ullrich collection.)

Charlotte (Buck) Oatman of Aurora was crowned 1964 queen by attorney at law Harvey Green. (J. F. Ullrich collection.)

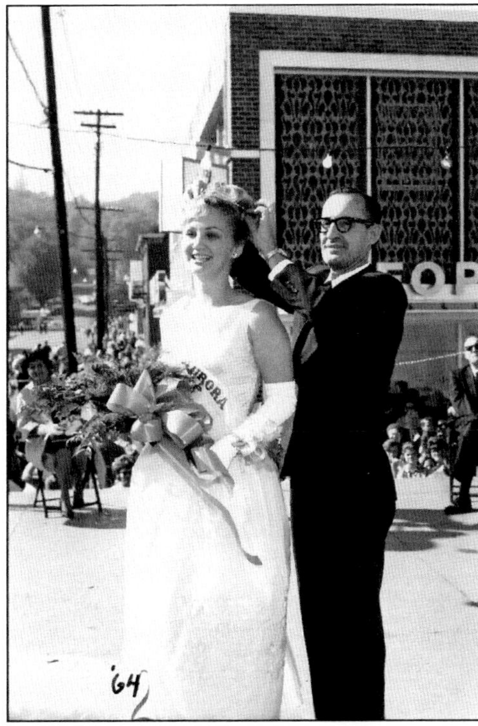

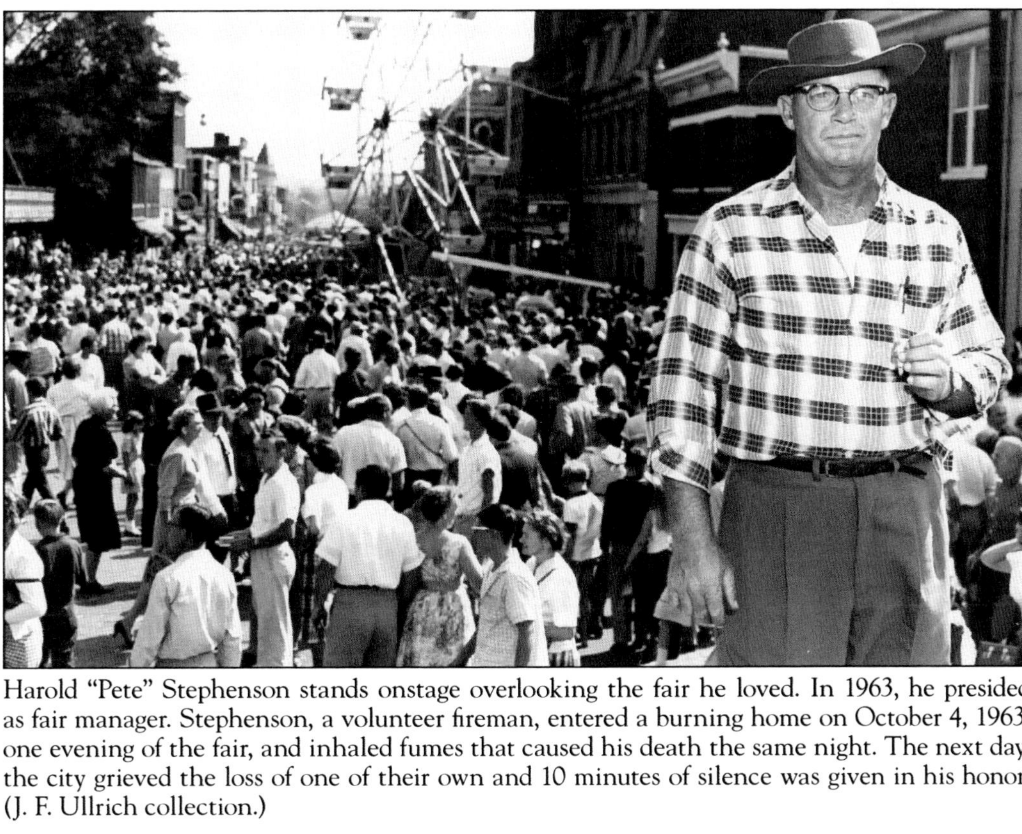

Harold "Pete" Stephenson stands onstage overlooking the fair he loved. In 1963, he presided as fair manager. Stephenson, a volunteer fireman, entered a burning home on October 4, 1963, one evening of the fair, and inhaled fumes that caused his death the same night. The next day, the city grieved the loss of one of their own and 10 minutes of silence was given in his honor. (J. F. Ullrich collection.)

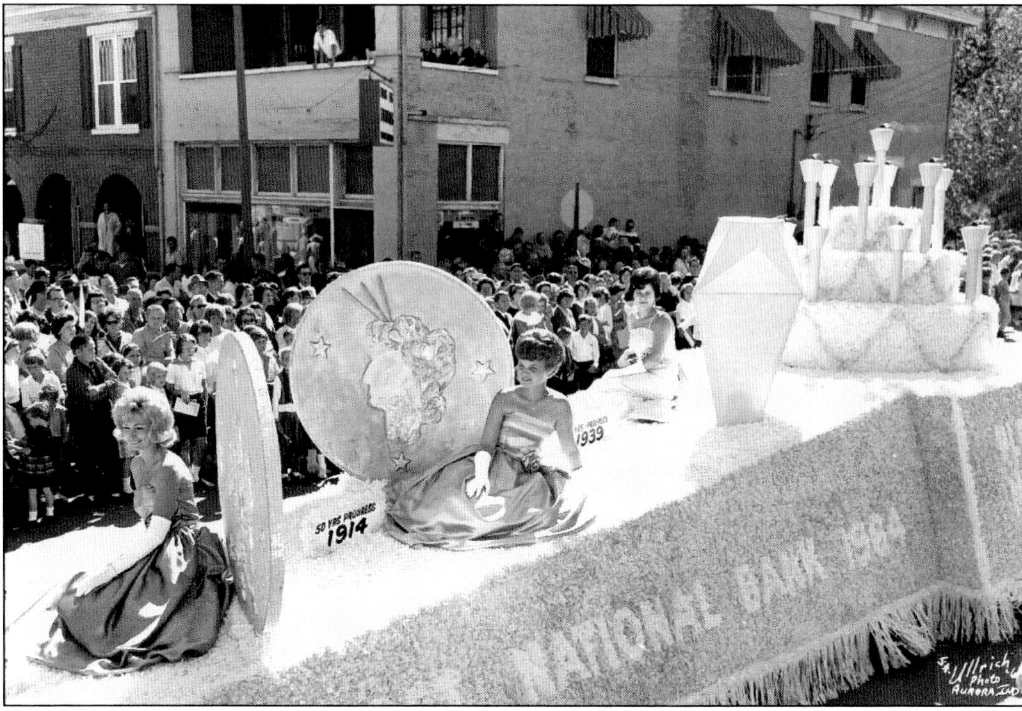

With the 1964 fair theme of Memories, the First National Bank netted two awards for best use of theme and the sweepstakes award. Sue Downey is seated center. (J. F. Ullrich collection.)

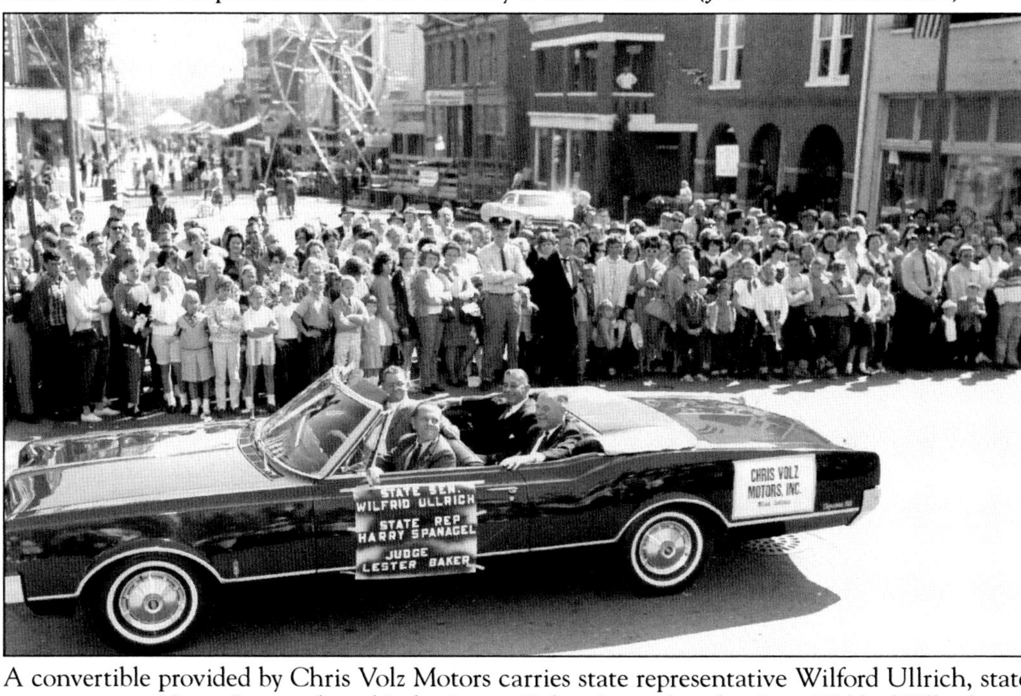

A convertible provided by Chris Volz Motors carries state representative Wilford Ullrich, state representative Harry Spanagel, and Judge Lester Baker. Later that day, Rep. Wilford Ullrich tied for first in the hog-calling contest and gave the pig to Charles Miller, a student at Aurora High School. (J. F. Ullrich collection.)

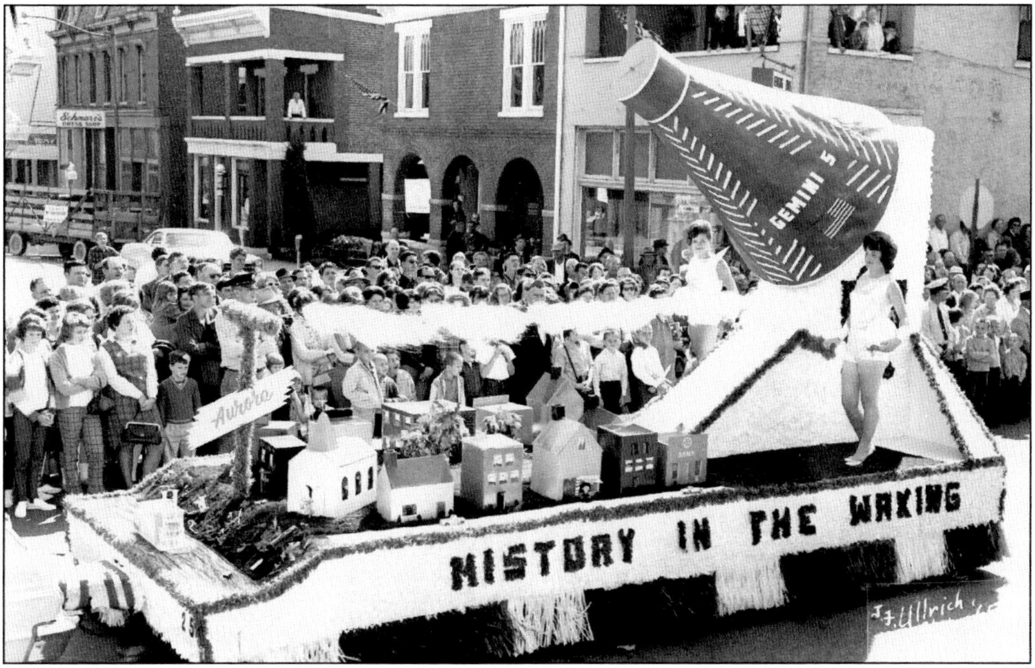

History in the Waking was the theme of the 1965 fair, and 44 floats entered the parade. Beta Sigma Phi won first prize in the contest with their entry of the Gemini 5 space capsule, orbiting over a model town of Aurora and the Ohio River. On the float are Jean Waite and Jean Molen as astronauts. (J. F. Ullrich collection.)

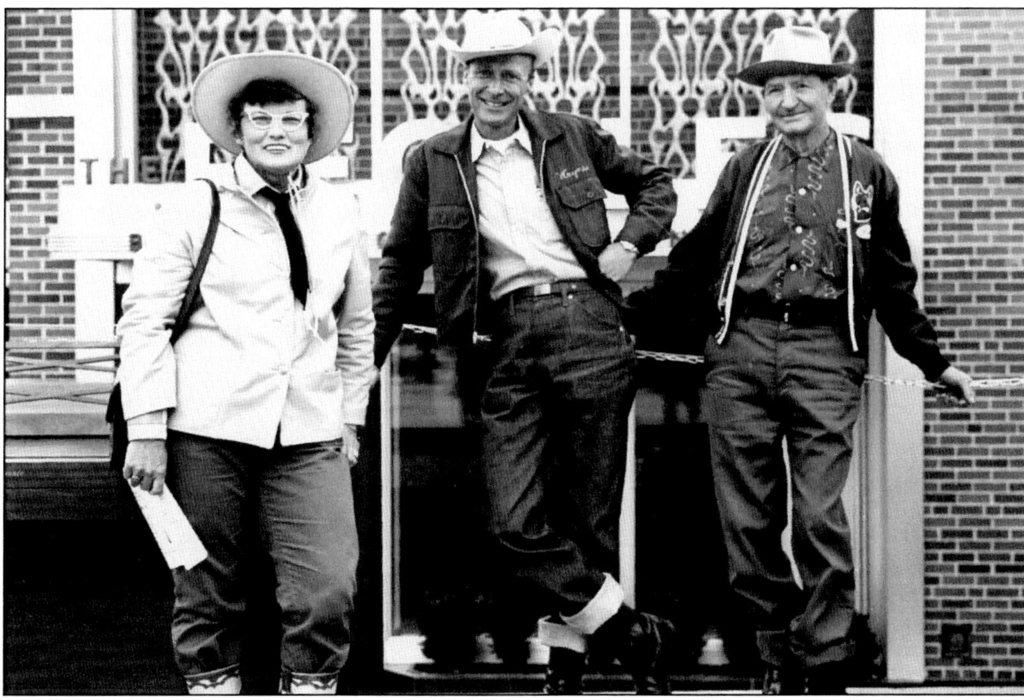

Seen here from left to right, pet parade judges Pody Lowe, Wayne Busse, and Amos Oberting casually wait onstage for the Friday afternoon event to begin. (J. F. Ullrich collection.)

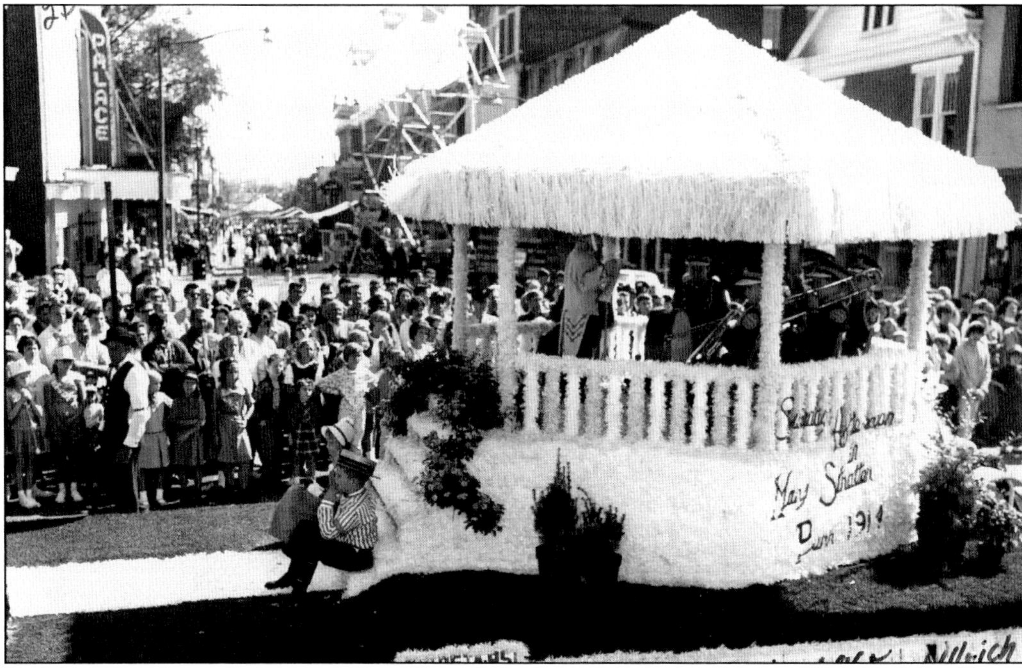

Once again a winner is the Phi Beta Psi float. The float is a replica of the bandstand erected in the Mary A. Stratton Park on Fifth Street in 1914. (J. F. Ullrich collection.)

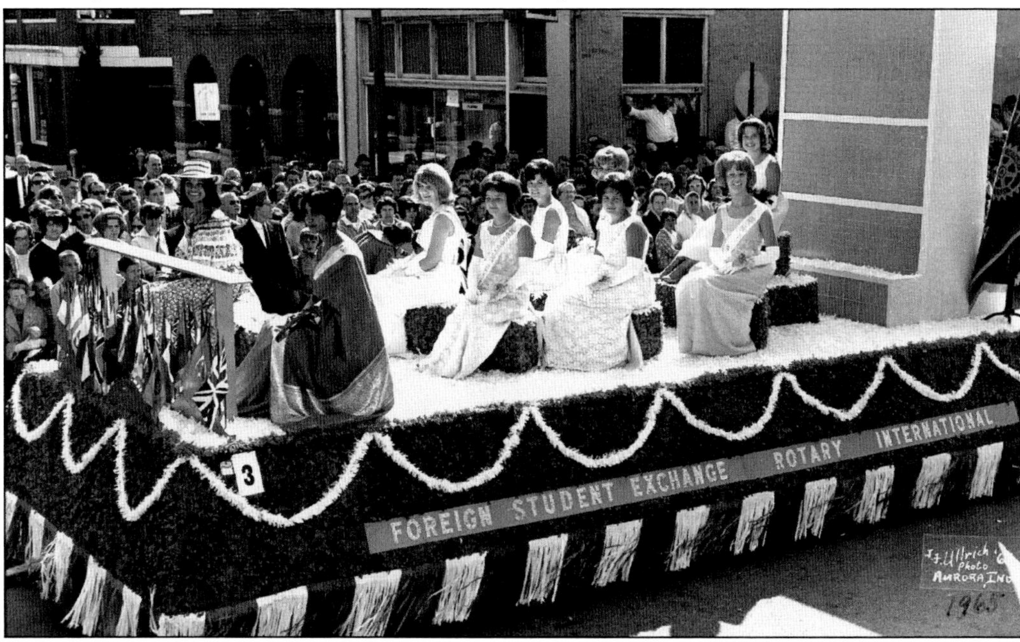

This float depicted the theme of International Understanding. Queen Sharon (Sterling) Rosenburgh, Aurora, is seen in 1965 with her court of Darlene Robinson, Patriot; Judith Kay Skidmore, North Dearborn; Sandra Kay Bentle, Moores Hill; Betty Cutter, Dillsboro; Linda Hiesman, Lawrenceburg; and Saundra Jo Moreillion, Rising Sun. The United Nations building is the backdrop. The Rotary Club also brought two international students from Western University. (J. F. Ullrich collection.)

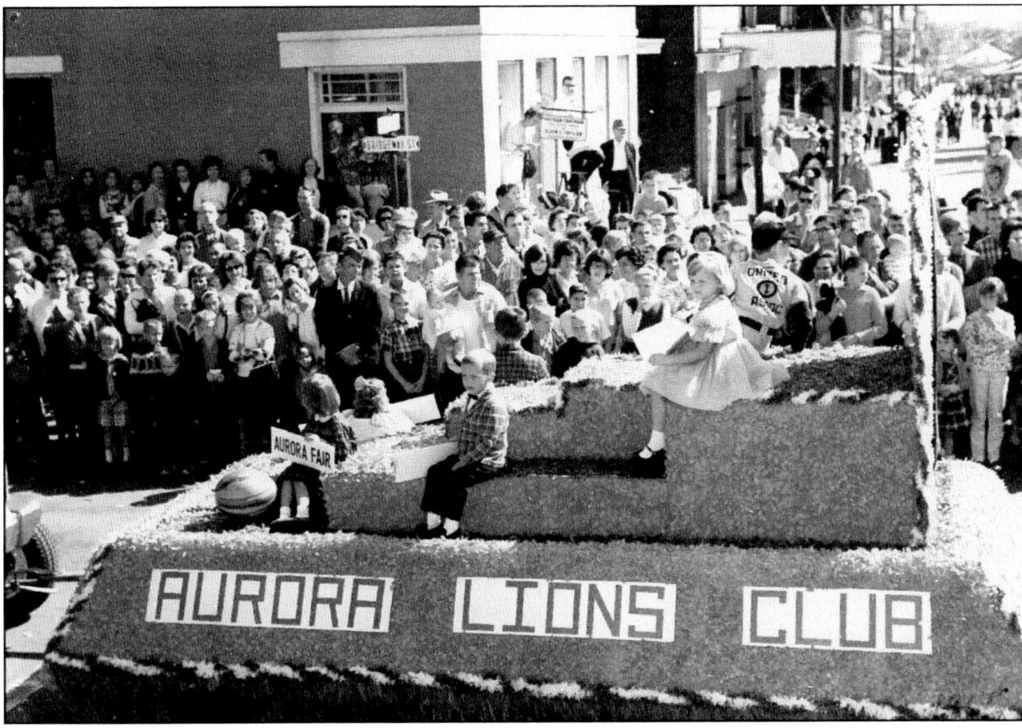

Small children represent the future of the farmers fair on the Lions Club float. (J. F. Ullrich collection.)

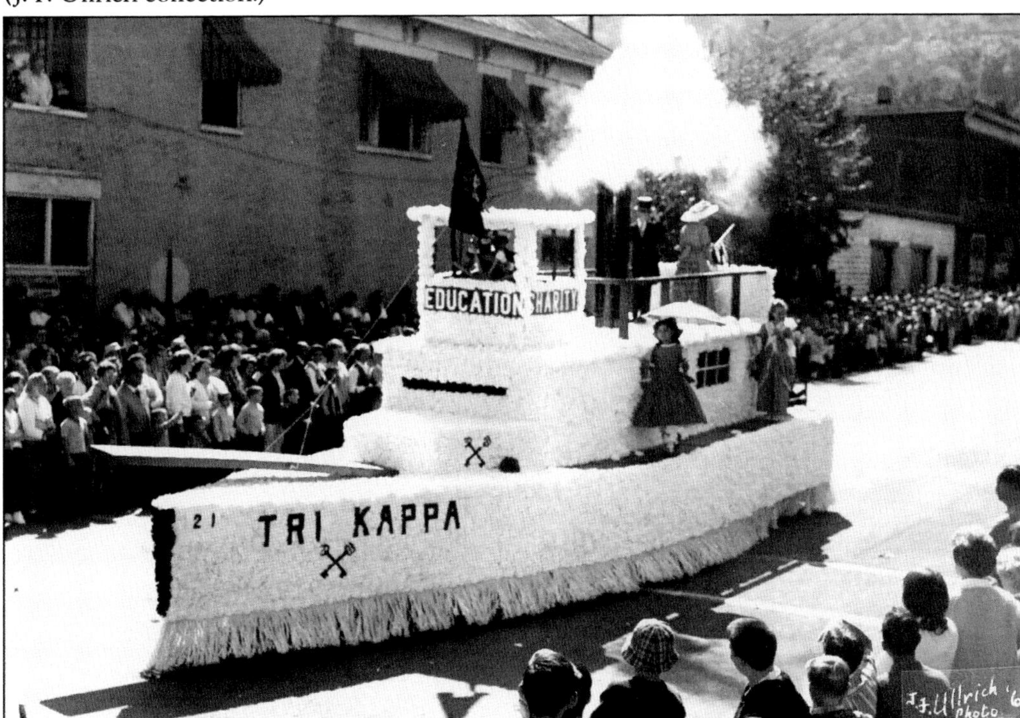

Seen here is the Tri Kappa Belle of Aurora float. (J. F. Ullrich collection.)

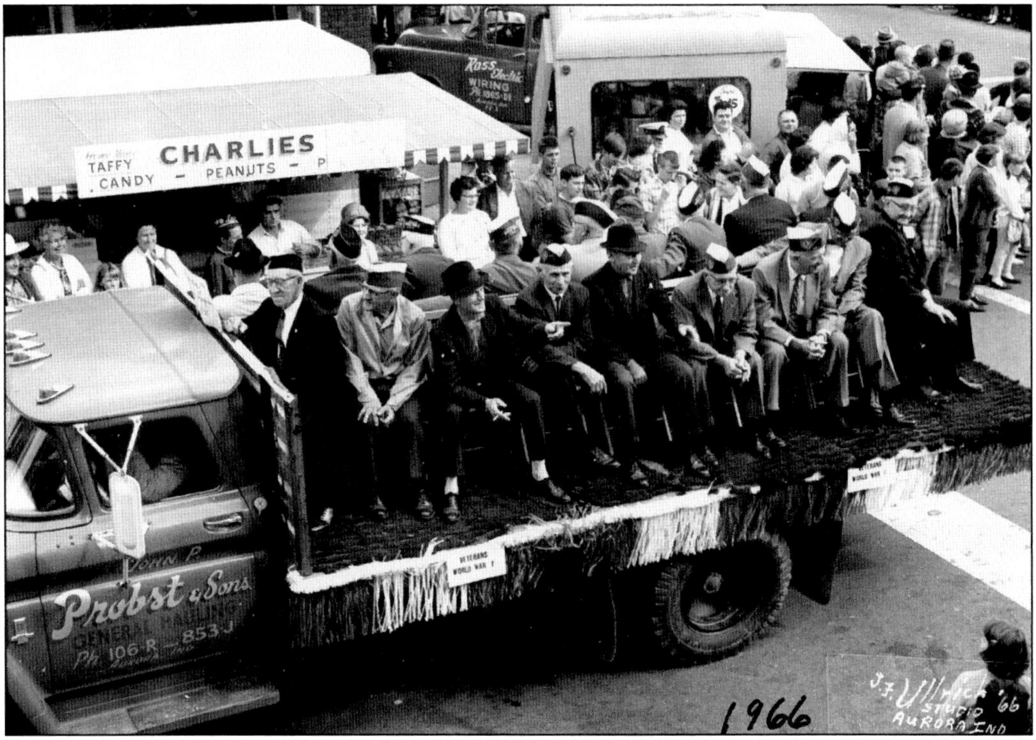

Earl Huffman, third from left, is seated along with his fellow World War I veterans. (J. F. Ullrich collection.)

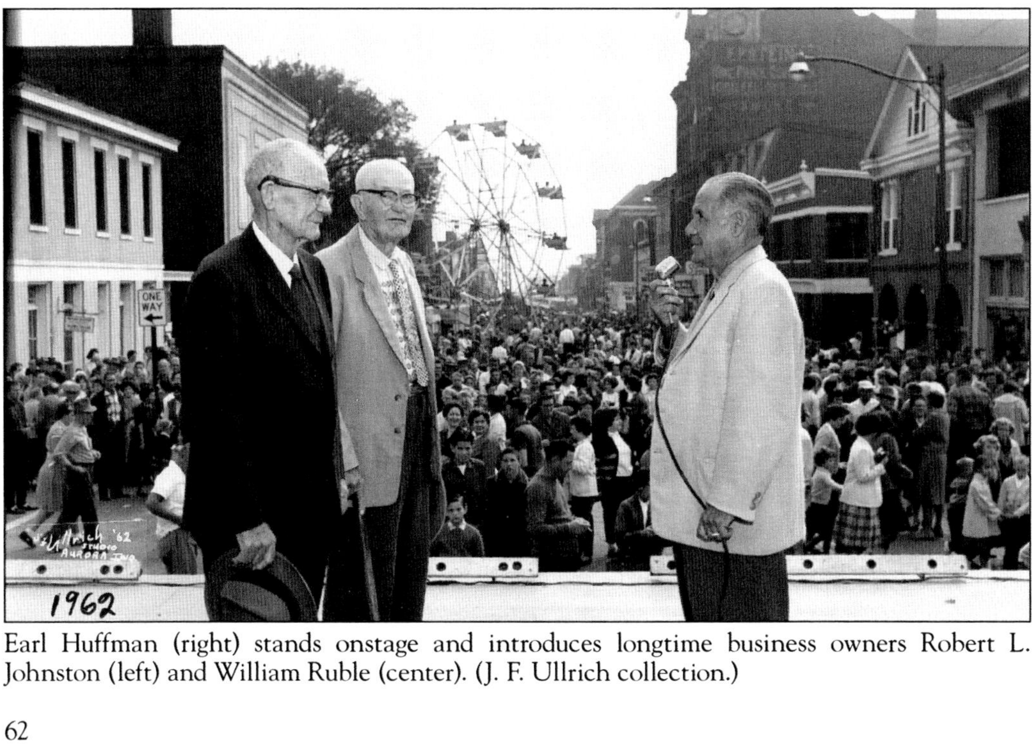

Earl Huffman (right) stands onstage and introduces longtime business owners Robert L. Johnston (left) and William Ruble (center). (J. F. Ullrich collection.)

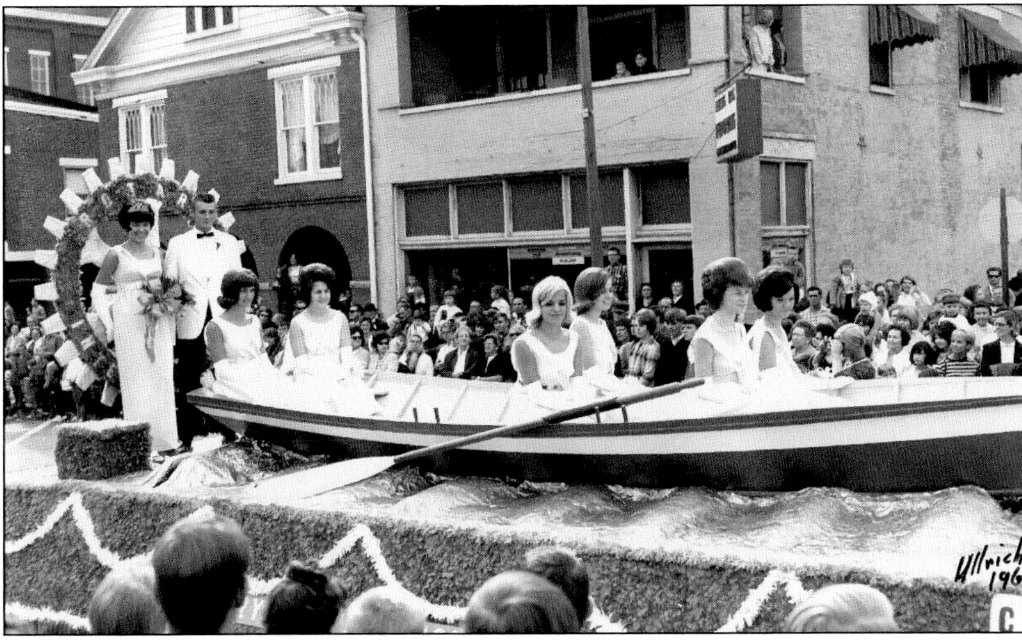

Queen Jackie Vail, Lawrenceburg, and King Harry Lyness, North Dearborn, stand at the rear of the float. Not in order are contestants Barbara Barrott, Aurora; Joyce Penn, Dillsboro; Bonnie Livingston, Moores Hill; Pam Fox, North Dearborn; Kathy Dick, Patriot; and Sondra Lou Courtney, Rising Sun. The year 1966 was the first time that young men were chosen to compete for king of the farmers fair. (J. F. Ullrich collection.)

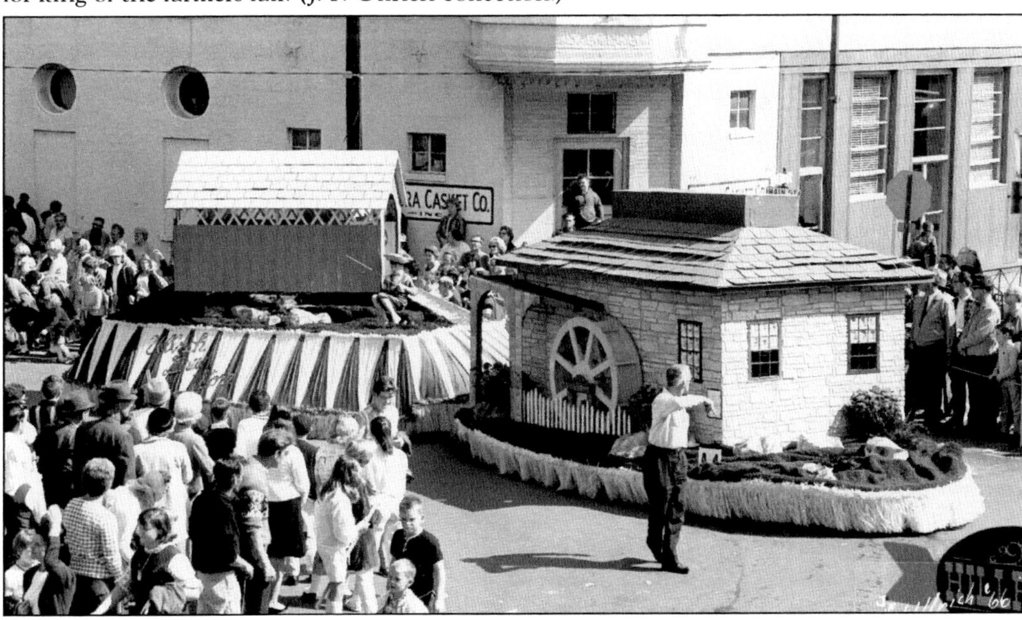

Gordon Britton guides the huge two-unit float around a crowded corner. One float displays a water wheel with running water turning the mill wheel and the other a landscape appropriate to the mill with a covered bridge and a running stream in which a boy was fishing for live fish. It was entered by Ullrich's Drug Store and was built by the Phi Beta Psi sorority. The float won the sweepstakes award for best depicting the parade theme, Reminiscing. (J. F. Ullrich collection.)

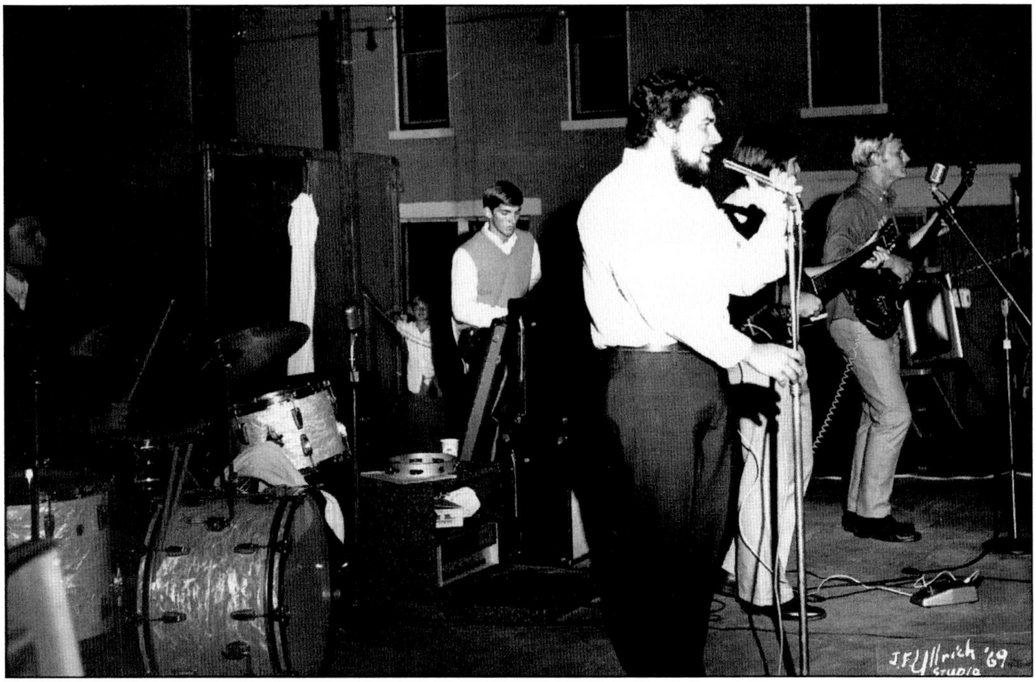

A favorite local band by the name the New Dukes entertains for the crowd. The band includes Nick Ullrich, front center; David Kling on the organ; Kenzie Bentle on drums; and Steve Fox and Willie Childers, guitar. (J. F. Ullrich collection.)

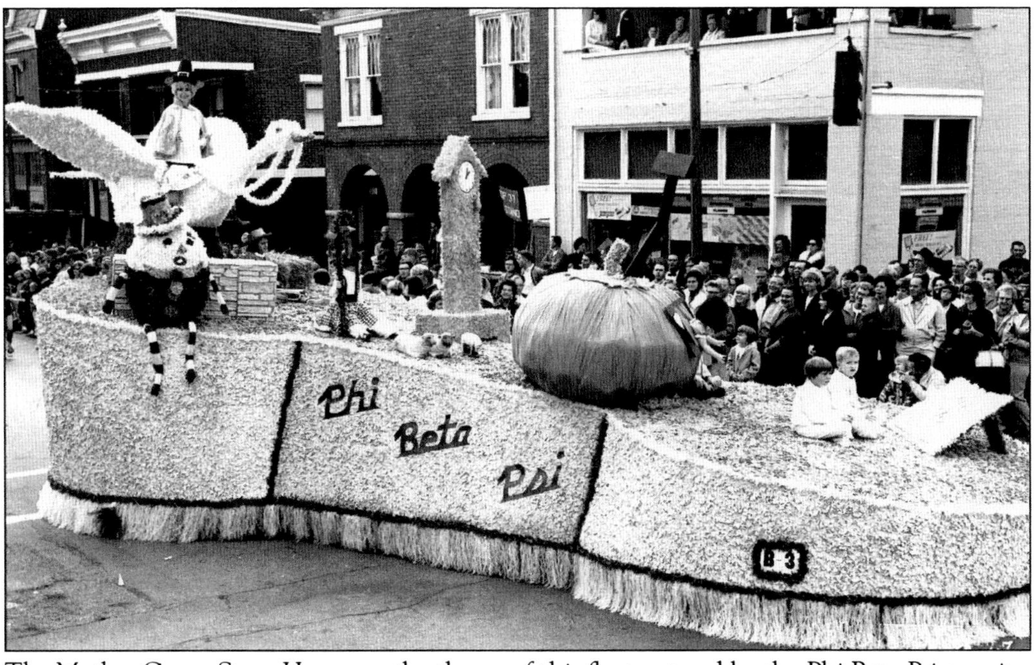

The Mother Goose Story Hour was the theme of this float entered by the Phi Beta Psi sorority. Mother Goose is depicted by Tracy Giffin. Lori Ann Lischkge and Scott Brown are reading the storybook while Anthony Meyer is portraying Peter Peter Pumpkin Eater. Mary Jane Teaney poses as Little Bo Peep, and Karen Harves portrays Little Boy Blue. (J. F. Ullrich collection.)

Mike Strasemeier, two and a half years old; Greg Slayback, one and a half years old; and Kim Slayback, four years old, enjoy a beautiful fall day riding a small army tank. (Strasemeier collection.)

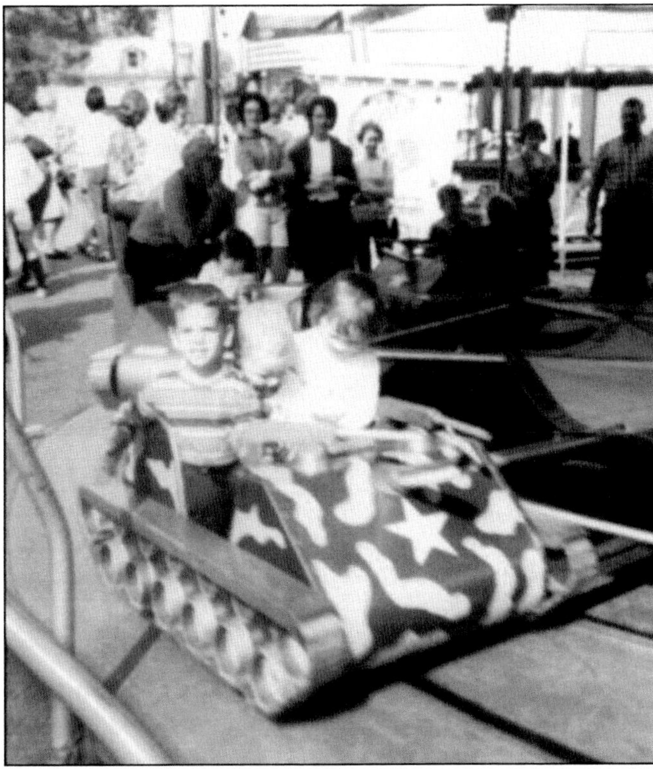

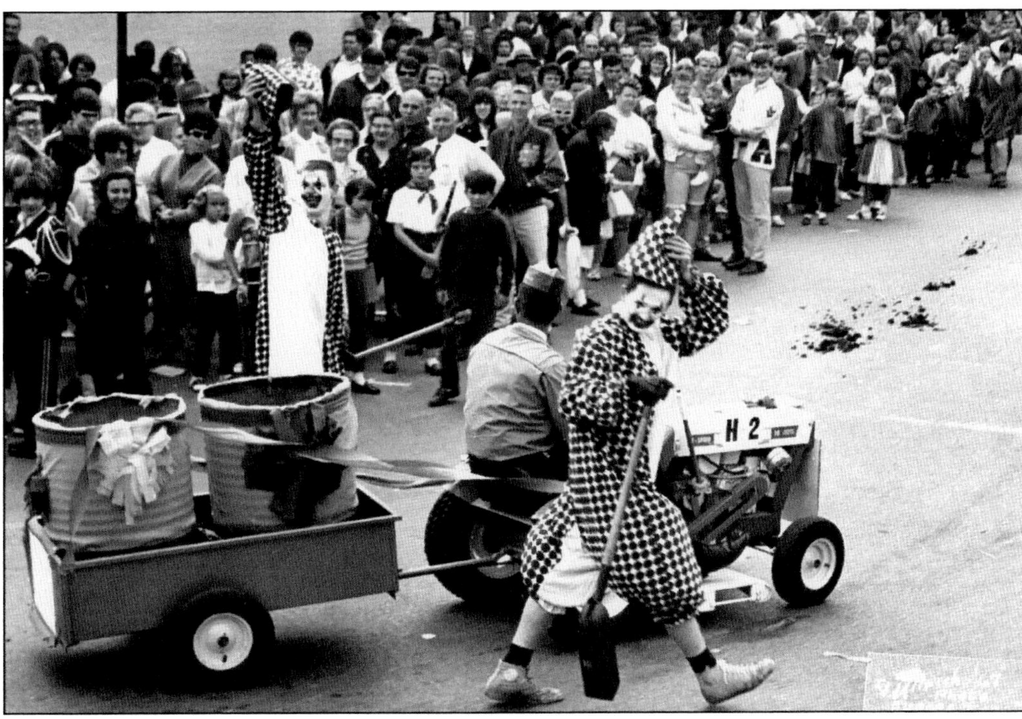

The crowd cheers for the clowns as they clean up the dirty work left by the horses. The clowns show their appreciation by taking a bow and tipping their hats. (J. F. Ullrich collection.)

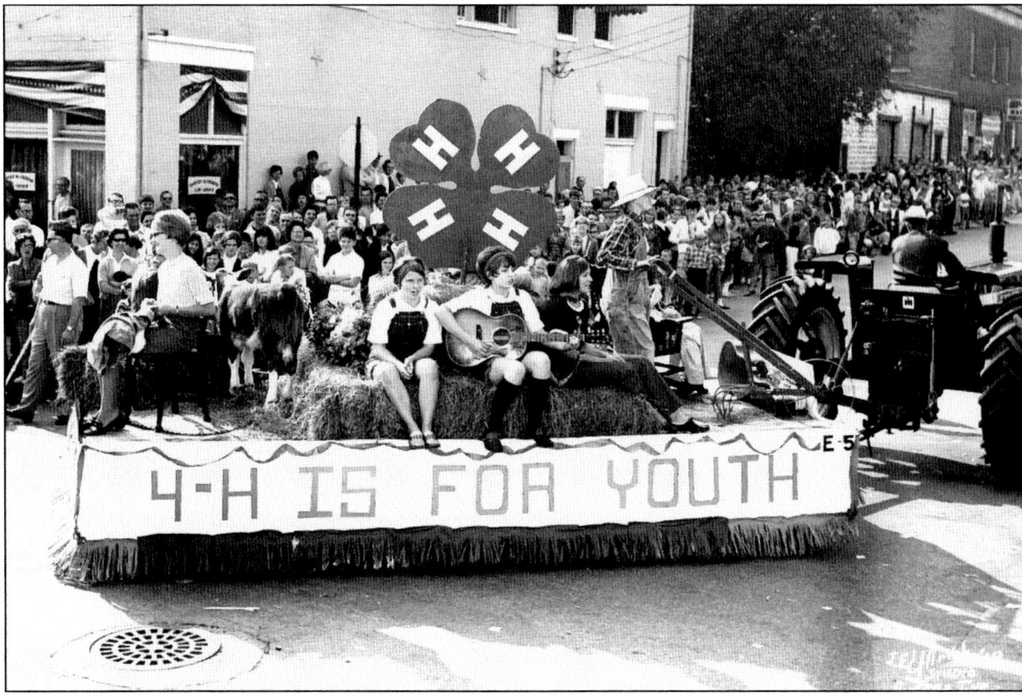

Area 4-H youth represent the various talents of the organization. A small calf calmly stands on the back of the float. (J. F. Ullrich collection.)

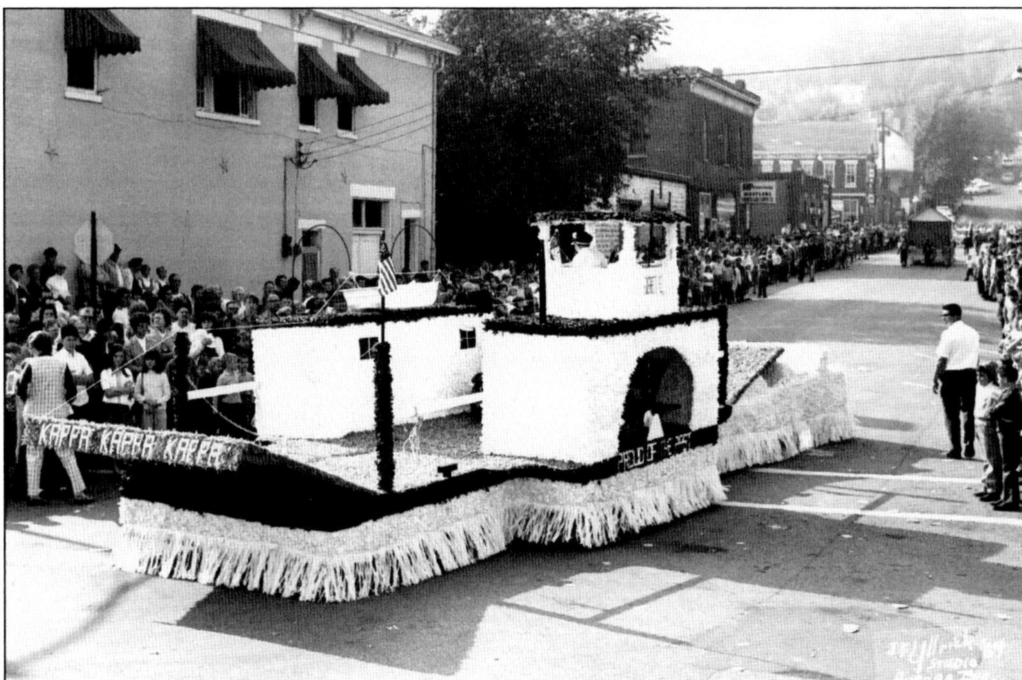

History in the Making is the theme of the 57th annual event. This float by the Tri Kappa sorority depicts the Aurora ferry. The ferry landing was located at the end of Third Street in Aurora. (J. F. Ullrich collection.)

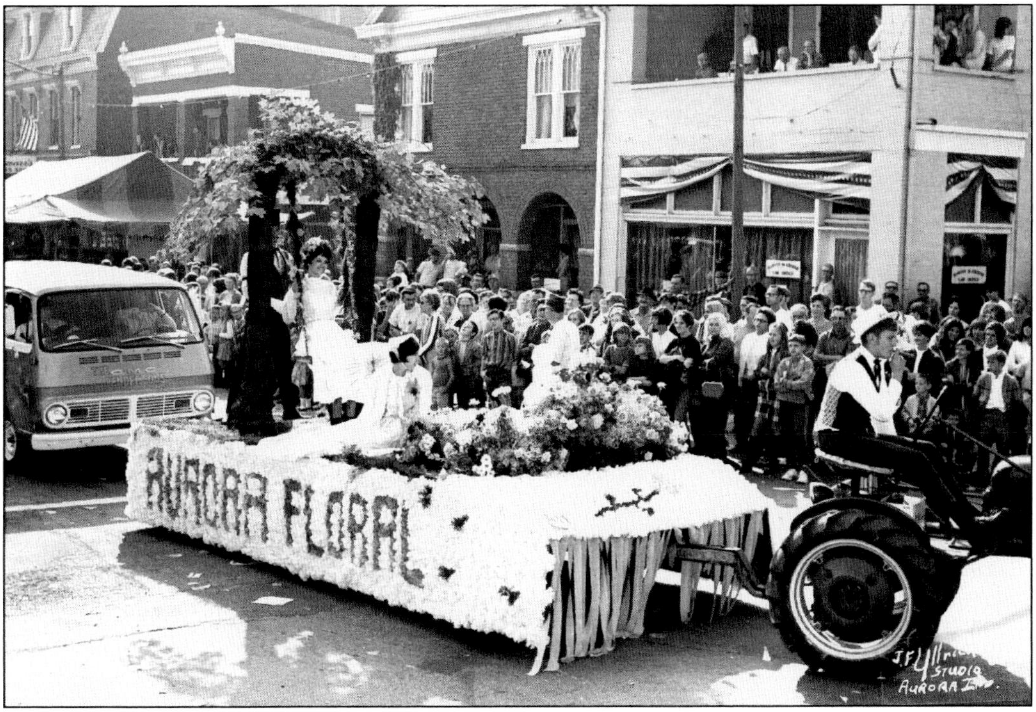

Kenny Peters gently pushes his wife, Zora Peters, on a swing. The float represented the Aurora Floral Shop. (J. F. Ullrich collection.)

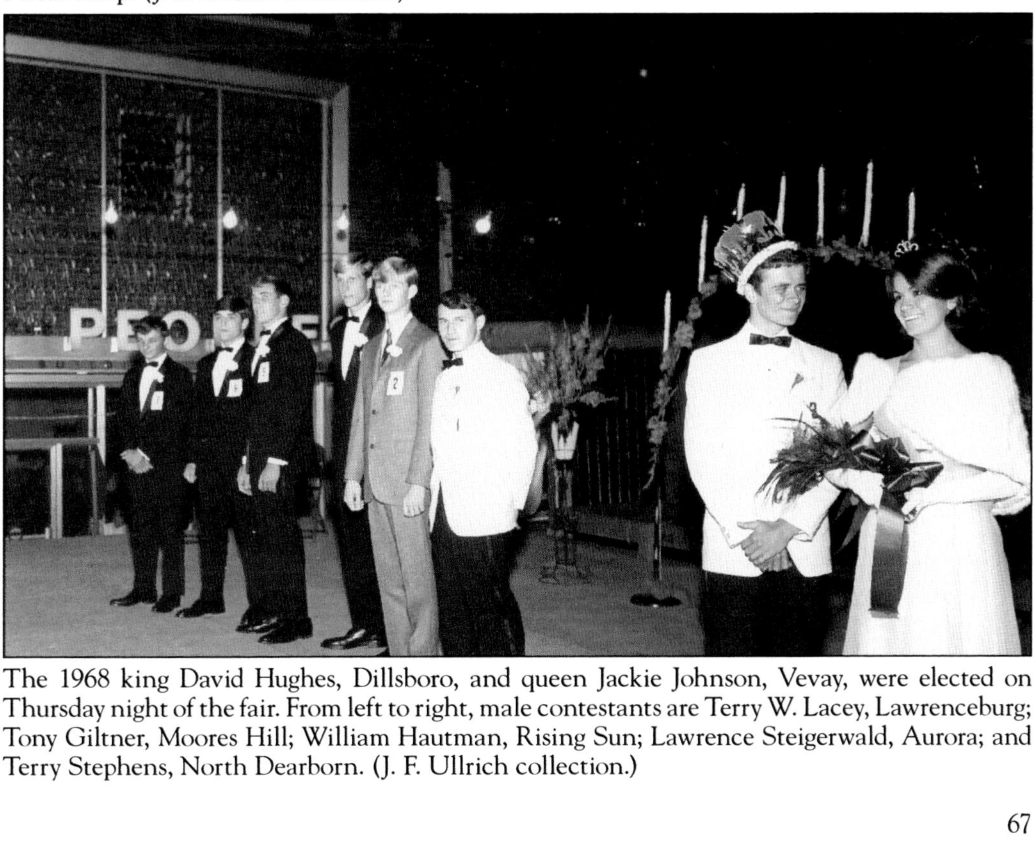

The 1968 king David Hughes, Dillsboro, and queen Jackie Johnson, Vevay, were elected on Thursday night of the fair. From left to right, male contestants are Terry W. Lacey, Lawrenceburg; Tony Giltner, Moores Hill; William Hautman, Rising Sun; Lawrence Steigerwald, Aurora; and Terry Stephens, North Dearborn. (J. F. Ullrich collection.)

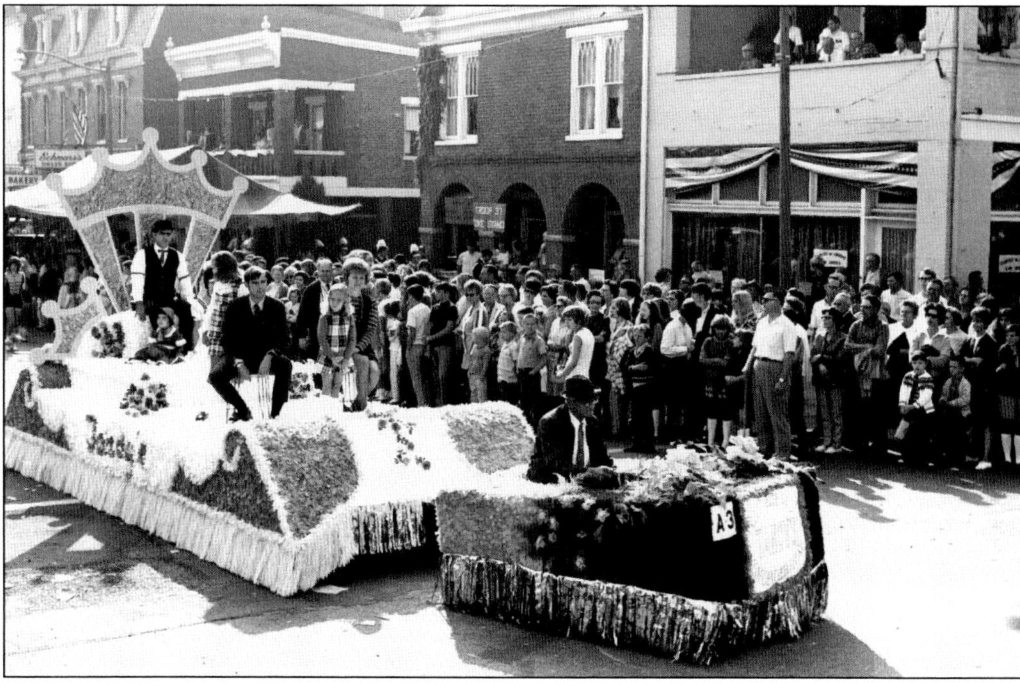

The Tandy family represents the Tandy's Clothing Store. Mike Swift is seated along with Dolores Tandy and her three daughters. Wood Whitaker is seated in the back, and Ervin Fillenwarth is driving. (J. F. Ullrich collection.)

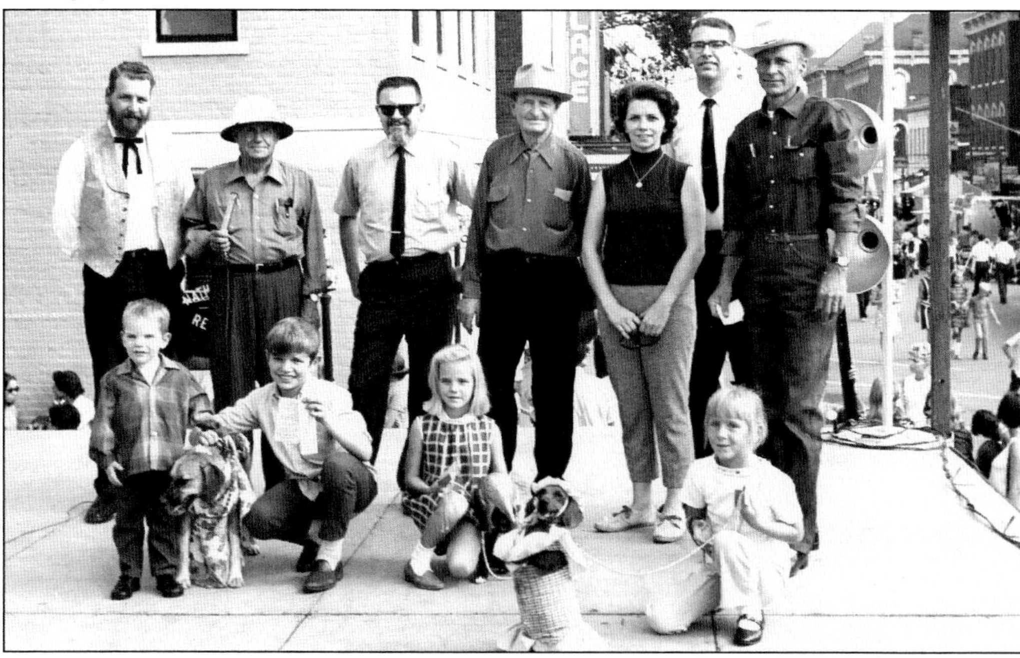

A dachshund hams it up, and only the best-dressed dog with a good personality can win this contest. The children in the first row are unidentified. Second row judges, from left to right, are Hubert Neff, Earl Huffman, Tony Lesko, Amos Oberting, Verna Smith, Dennis Meister, and Wayne Busse. (J. F. Ullrich collection.)

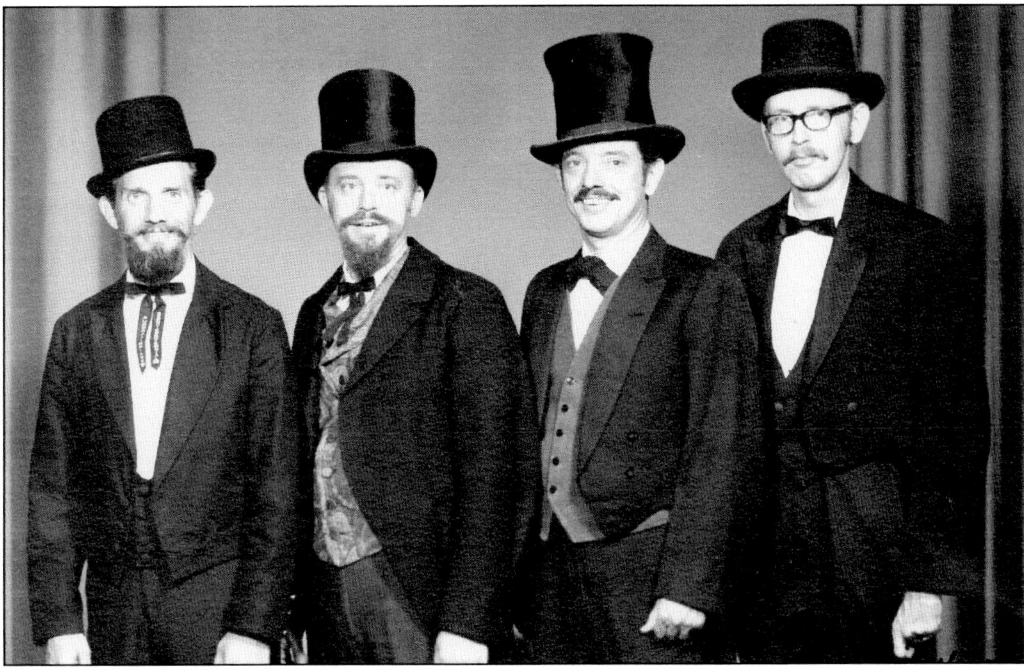

Mayor Bill Spencer appointed the 1969 Aurora sesquicentennial committee. Members, from left to right, are Kent Molen, president; Gary Meyer, vice president; John Sutton, treasurer; and Jim Randall, secretary. In the spirit of the event, the committee grew facial hair and dressed in period costumes. The most outstanding beard award went to Richard Lattire (not shown). (J. F. Ullrich collection.)

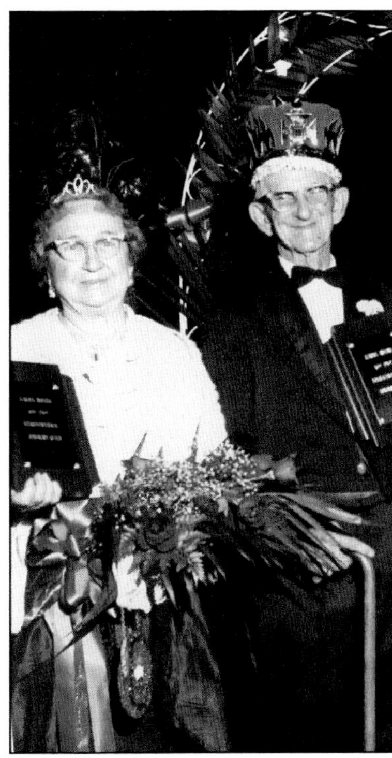

Daisy Jackson, voted queen of the Aurora sesquicentennial by the public, and king August Baer make a handsome pair on the stage. They were two of several candidates voted on by the public. The Aurora Journal-Bulletin sponsored the project. (J. F. Ullrich collection.)

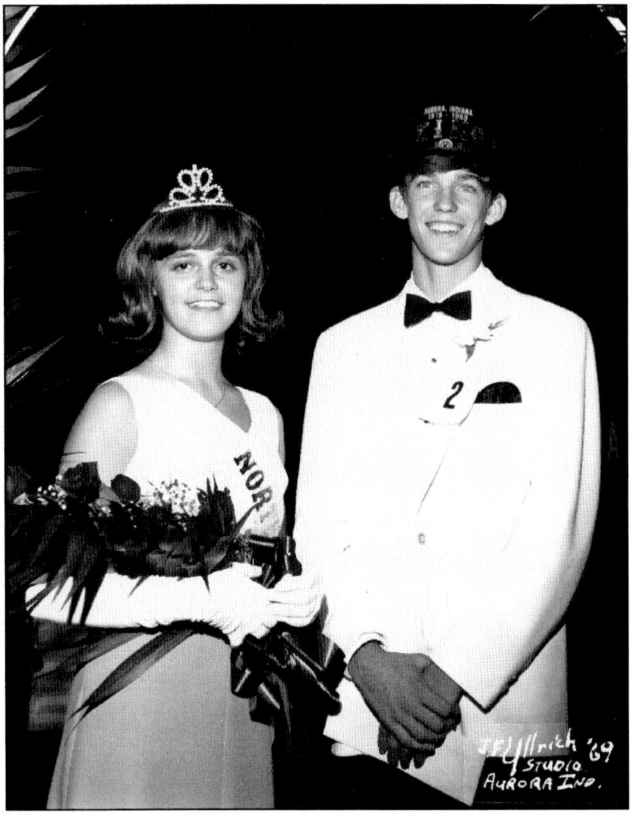

The 1969 queen Vicki Hester of North Dearborn and king Eddie Edwards of Aurora are all smiles for the camera after they received their crowns. (J. F. Ullrich collection.)

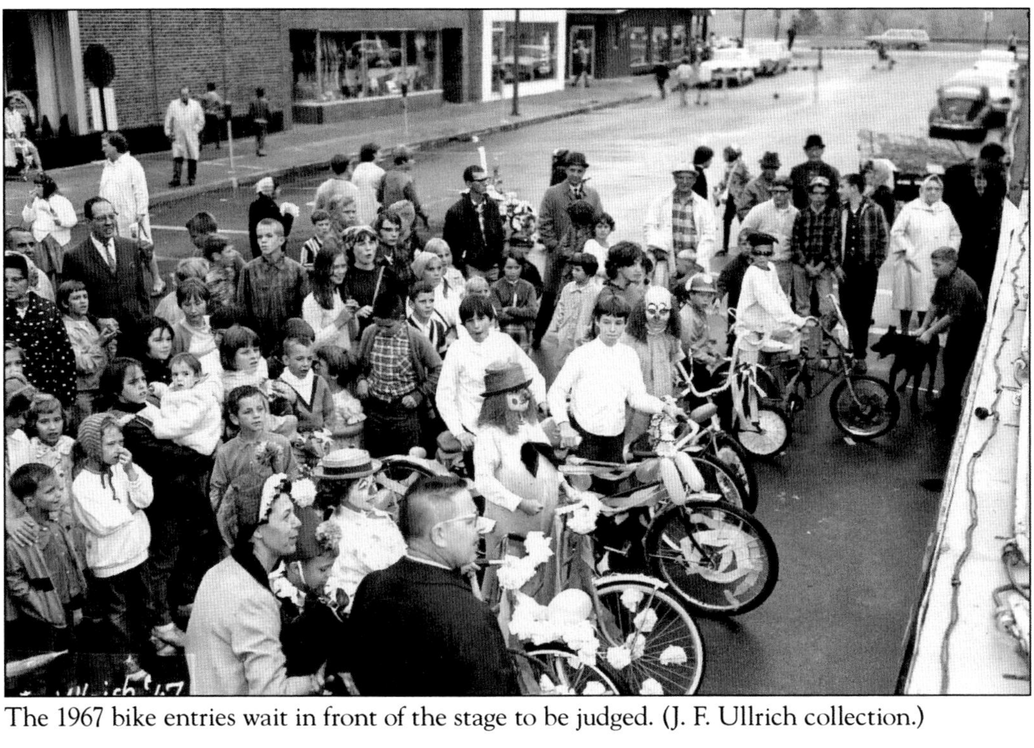

The 1967 bike entries wait in front of the stage to be judged. (J. F. Ullrich collection.)

Four

From 1970 to 1989

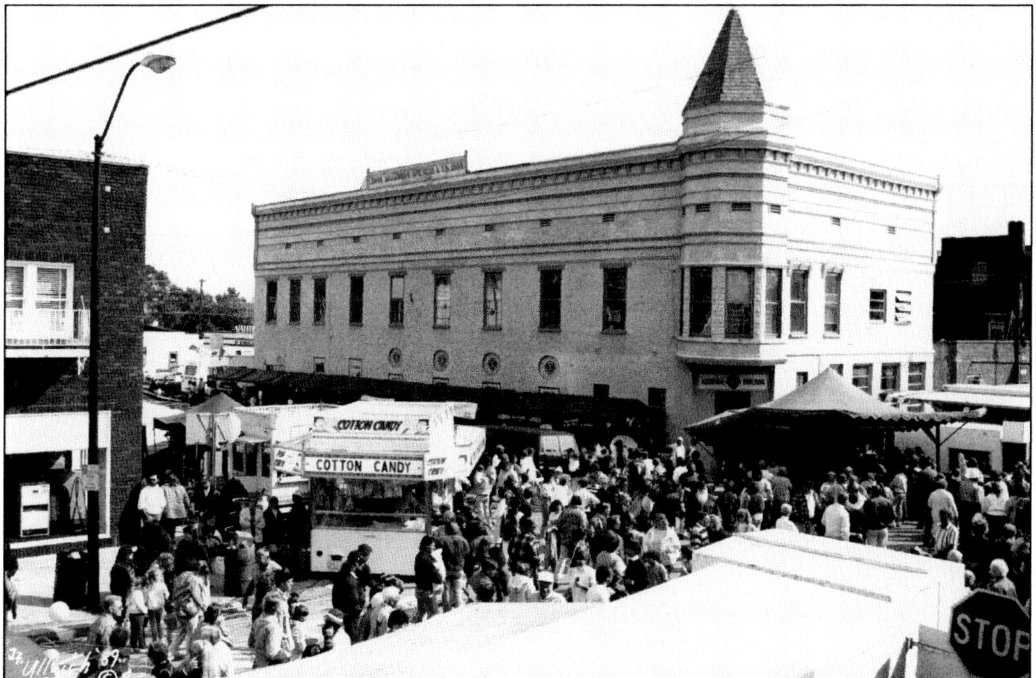

At one time, the Chamber-Stevens building stood not just as a backdrop for the Aurora Farmers Fair but as a place for the Lions Club to display the various fair exhibits. It was built on the corner of Main and Second Streets in 1840 and remodeled in the early 1900s. The old building fit the needs of the Lions Club. The space was utilized for many purposes, including community use for meetings, large gatherings, and farmers fair exhibits. In 1998, the building was totally destroyed by fire. Although years of memorabilia could not be replaced, the Lions Club members pulled together and rebuilt a similar building on the same corner. (J. F. Ullrich collection.)

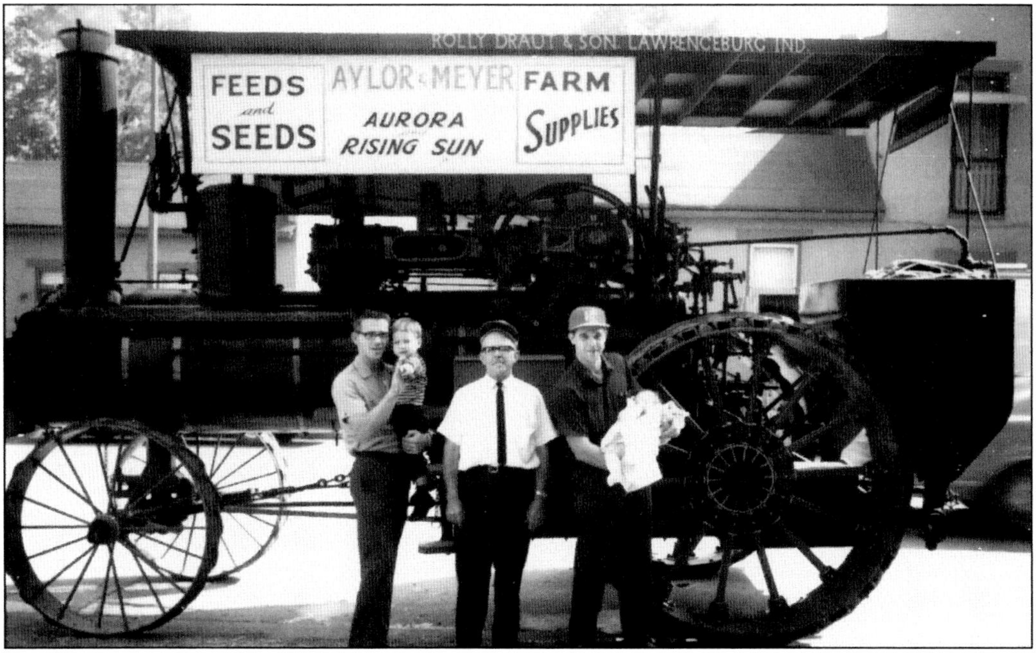

Howard Aylor and his family stand in front of a rare old steam engine. Steam engines were used at the end of the 1800s to the early 1900s. Many farmers could not afford the steam engine, so they hired threshing contractors. The men that owned their own engine and thresher went to different farms, hiring themselves out to thresh grain. (Aylor collection.)

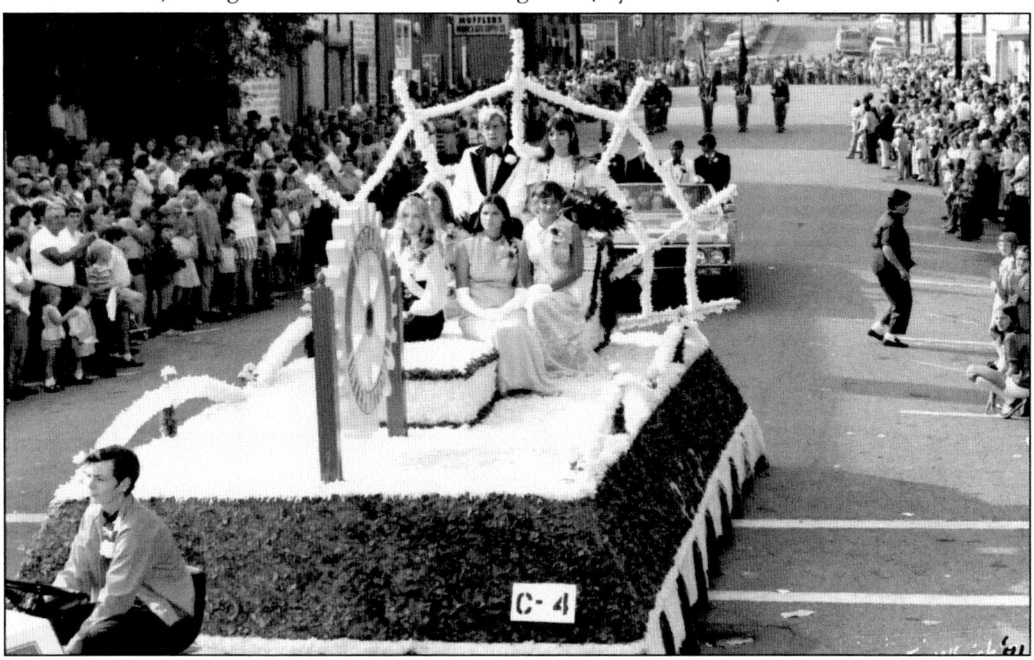

Susan Lewis of Dillsboro and Steve Foley of Moores Hill reigned over the 1971 farmers fair. The queen was sponsored by the Aurora Rotary Club and the king by the Dearborn County Register. James Brown, Rotary president, crowned the queen, who in turn crowned the king. (J. F. Ullrich collection.)

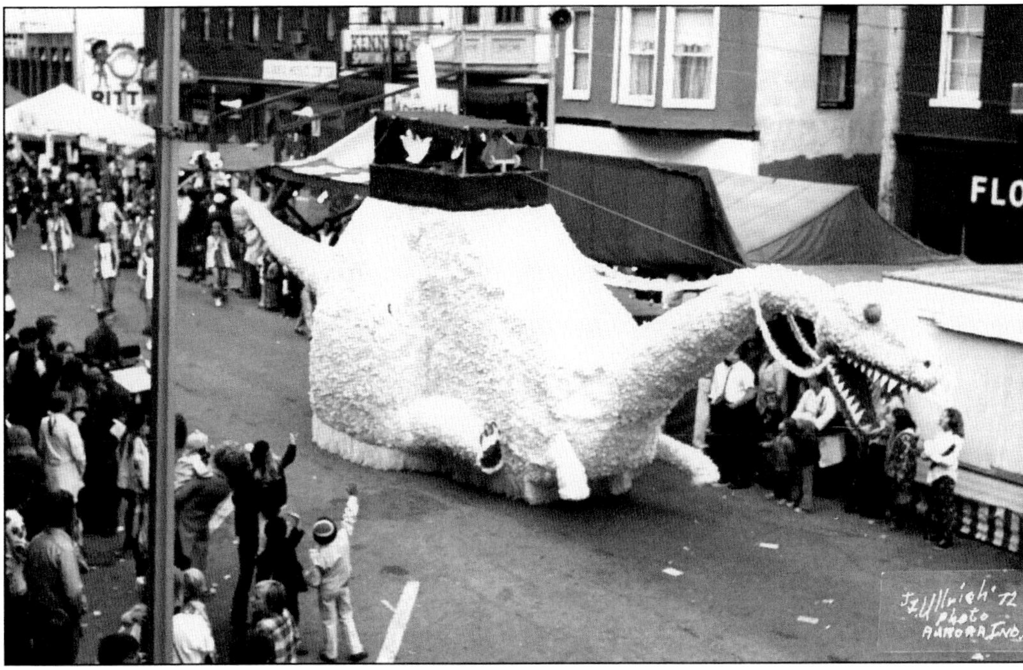

The Flintstone's giant dinosaur was built by the Aurora High School industrial arts class and was entered in the parade in 1972. (J. F. Ullrich collection.)

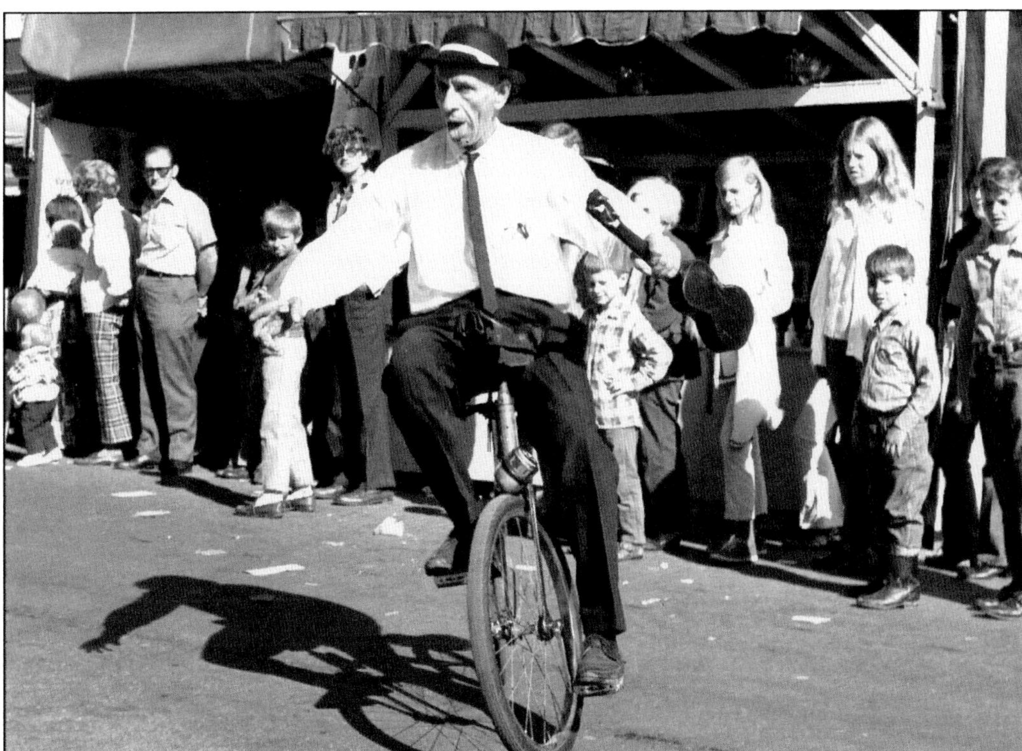

A team of unicycle riders entertained the crowd during the parade in 1970. (R. Ullrich collection.)

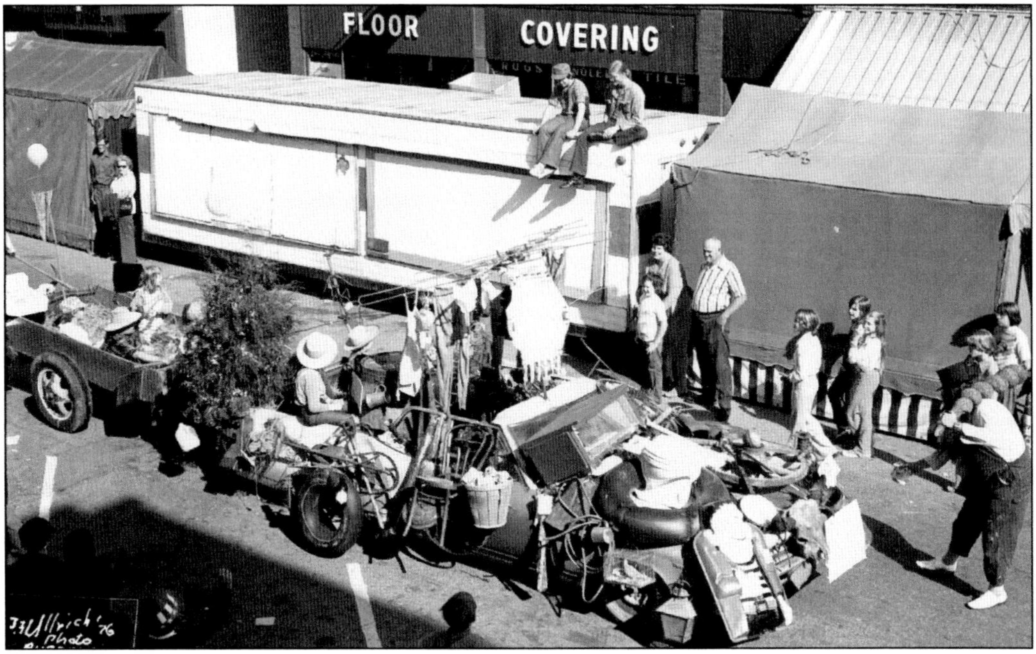

This 1976 humorous entry by the Linkmeyer brothers took first prize. The mischievous brothers "borrowed" a car from Gilstrap Motors, Aurora, and cut the top off. They loaded it up with everything from pigs to grandpa, grandma, and the grandkids. After the parade they gave the car back to a stunned Gilstrap. (J. F. Ullrich collection.)

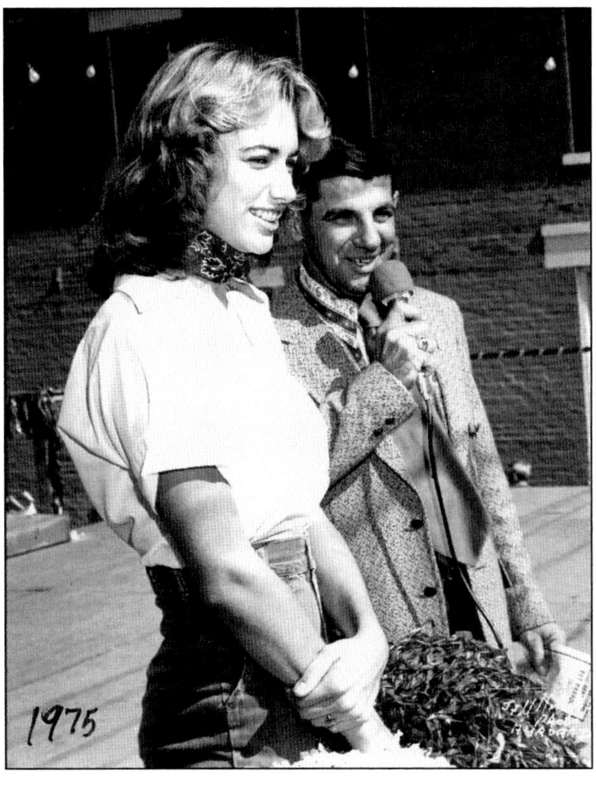

During the pet parade, Cindy Rullman is called onstage by Gene Weaver for an interview. Rullman was elected the 1975 queen. Rullman's parents, Millard and Sally, both owned businesses in Aurora. (J. F. Ullrich collection.)

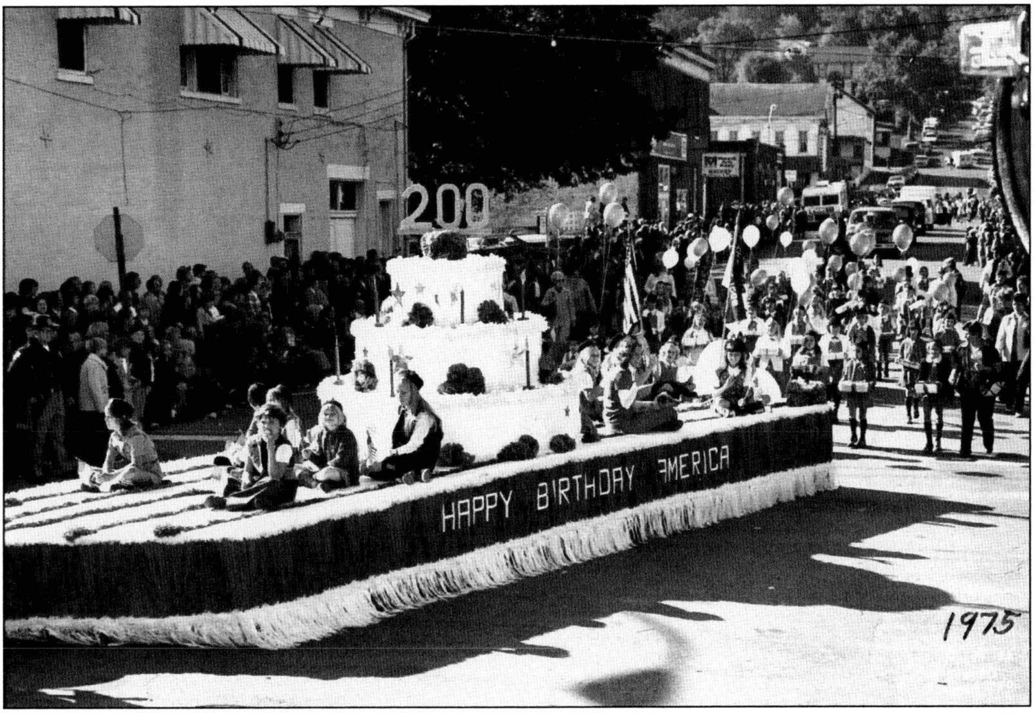

With the bicentennial around the corner, the Lions Club chose a patriotic theme for the 1975 fair. The Girl Scouts entered a Happy Birthday America float. (J. F. Ullrich collection.)

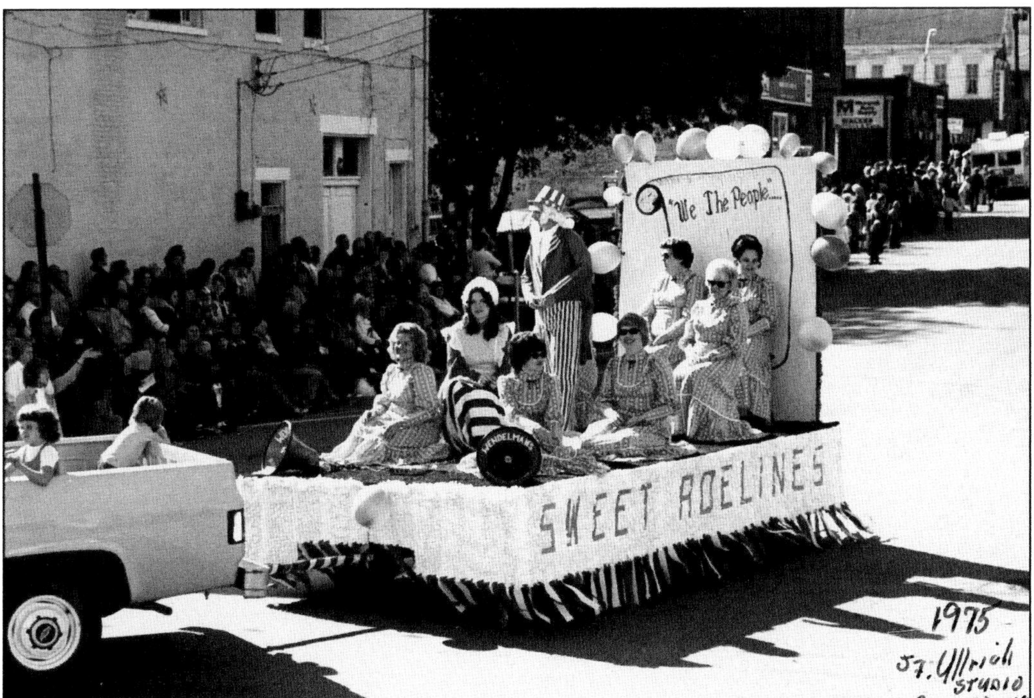

Uncle Sam rides a patriotic float with the singing group Sweet Adelines. (J. F. Ullrich collection.)

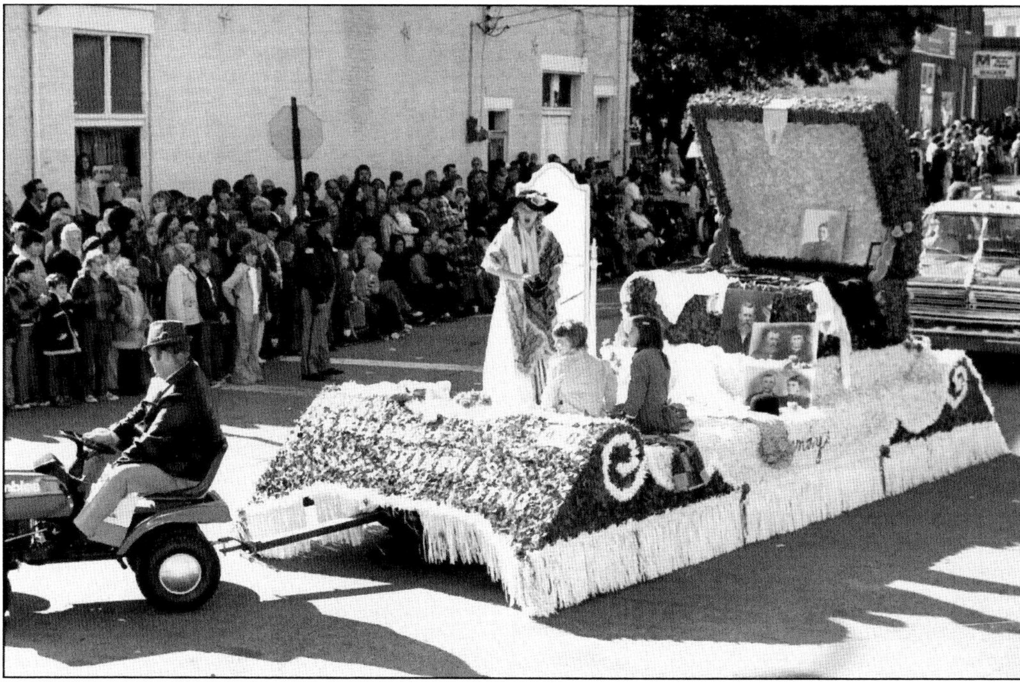

Children play pretend dress-up with their grandmother's clothing on the Tandy's Clothing Store float. Owner Jack Tandy is pulling the float, standing is Kathy Tandy, and seated are Chris Miller and Cindy Glenn. (J. F. Ullrich collection.)

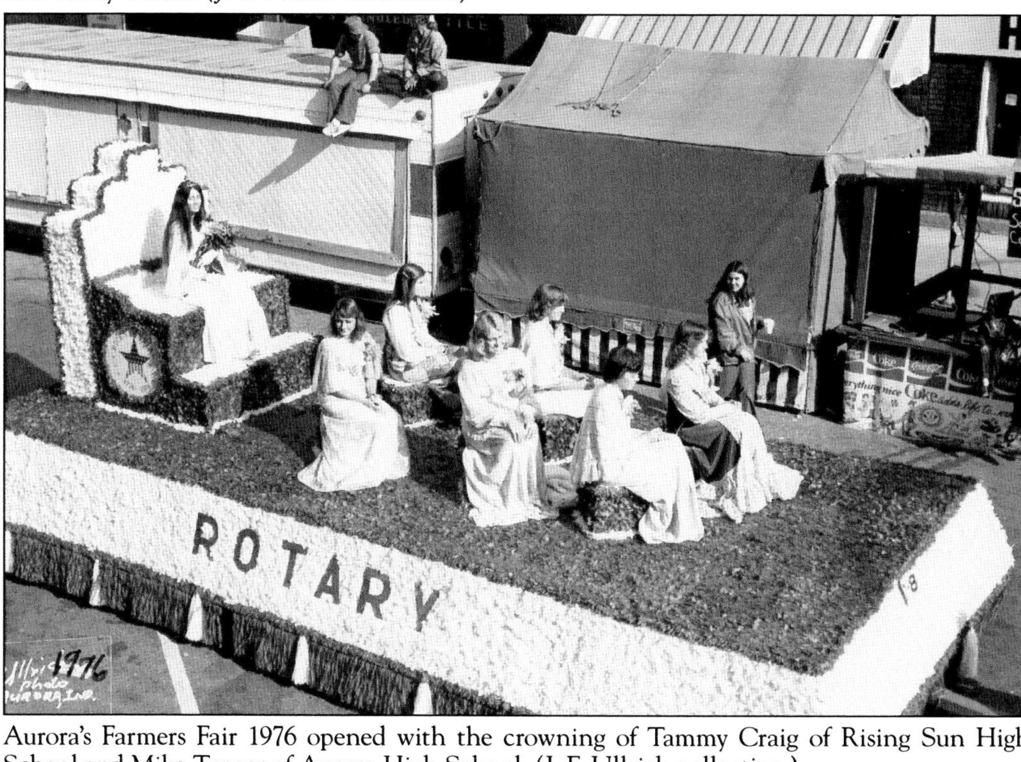

Aurora's Farmers Fair 1976 opened with the crowning of Tammy Craig of Rising Sun High School and Mike Turner of Aurora High School. (J. F. Ullrich collection.)

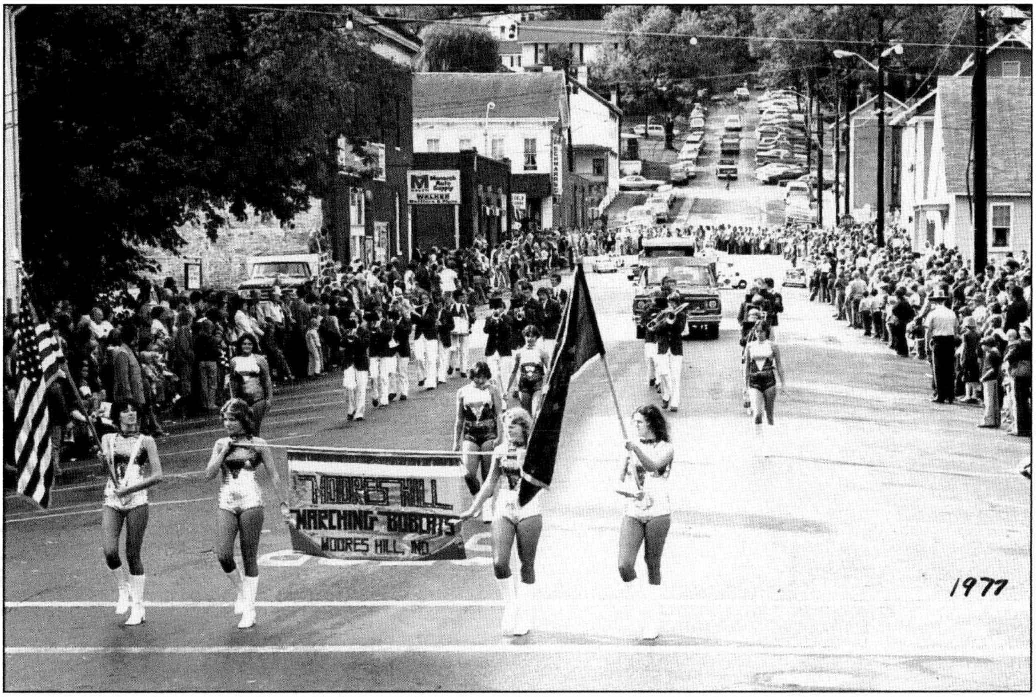

Marching in front of the stage is the 1977 Moores Hill High School band. Moores Hill Schools eventually consolidated with Dillsboro and Aurora School systems to form the South Dearbon School Corporation. (J. F. Ullrich collection.)

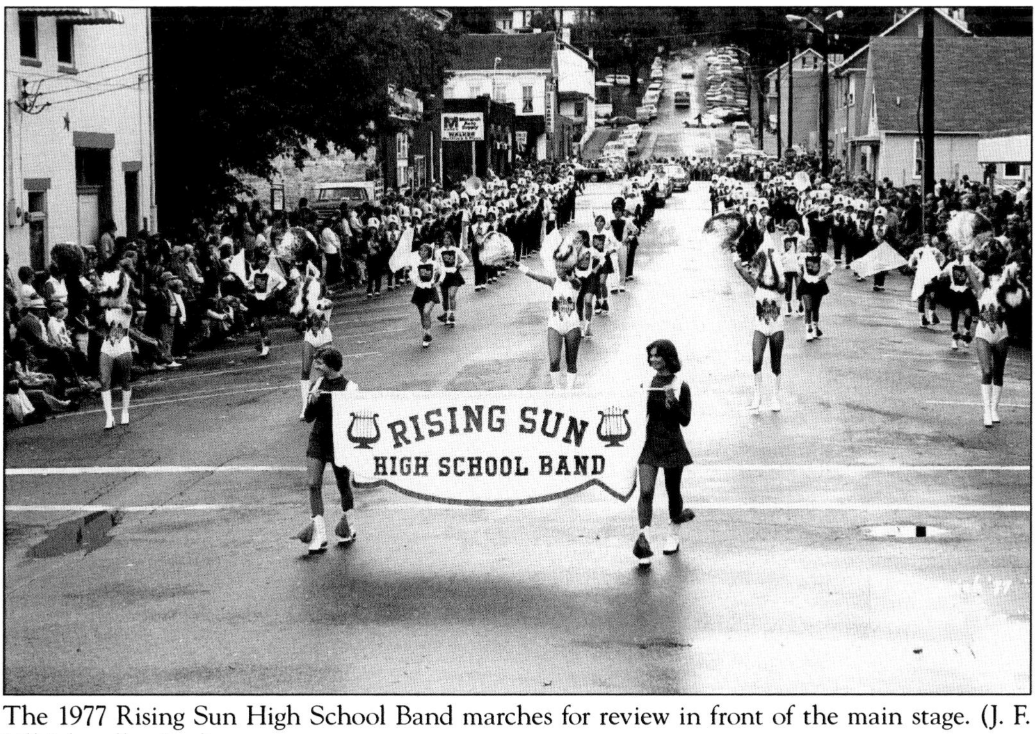

The 1977 Rising Sun High School Band marches for review in front of the main stage. (J. F. Ullrich collection.)

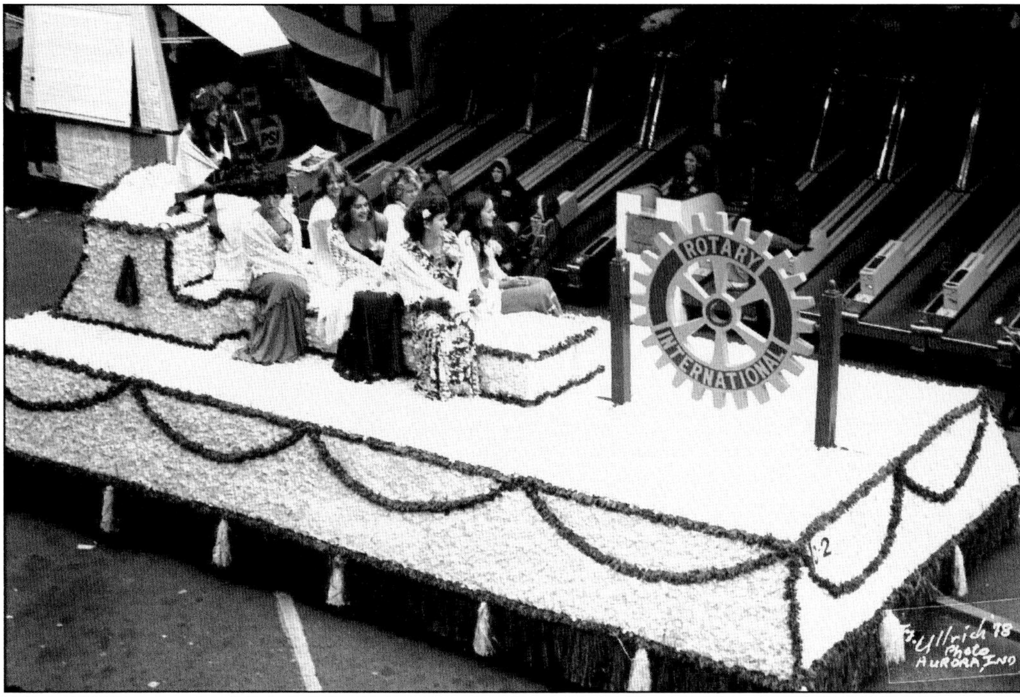

The 1978 queen and her court are sitting on the Rotary float. Malcolm Markland of Switzerland County was crowned king, and Donna Speer from South Ripley reigned as queen. Speer was honored as queen of the Versailles Pumpkin show the week before. (J. F. Ullrich collection.)

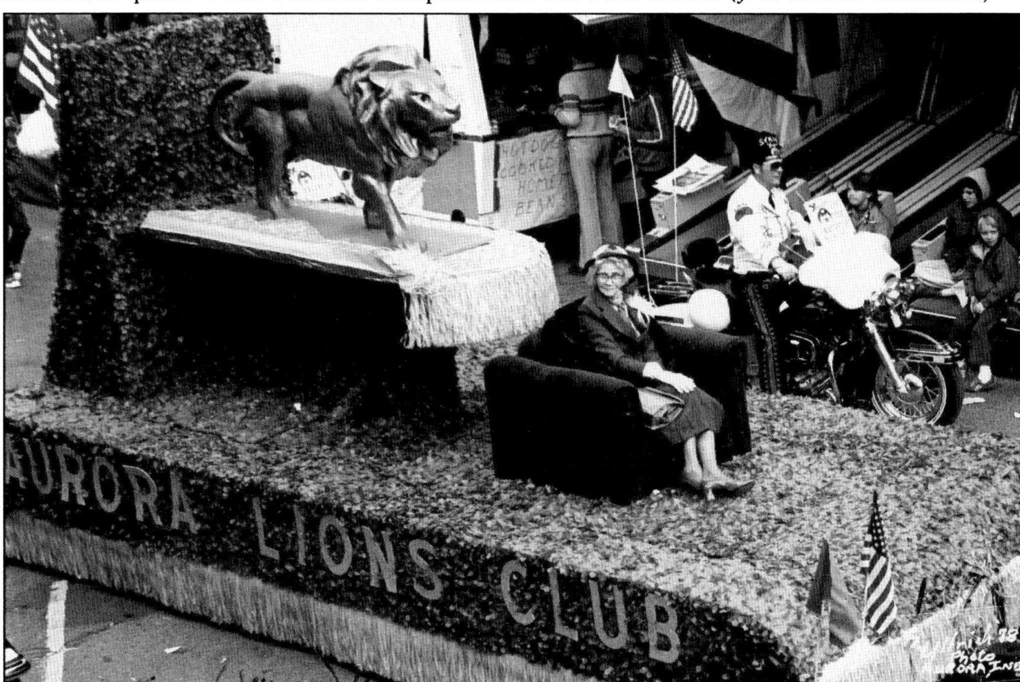

Grand marshal Ester (Wilson) Roache is the daughter of the original fair organizer, C. B. Wilson. Roache was an active member of the Hillforest Mansion. (J. F. Ullrich collection.)

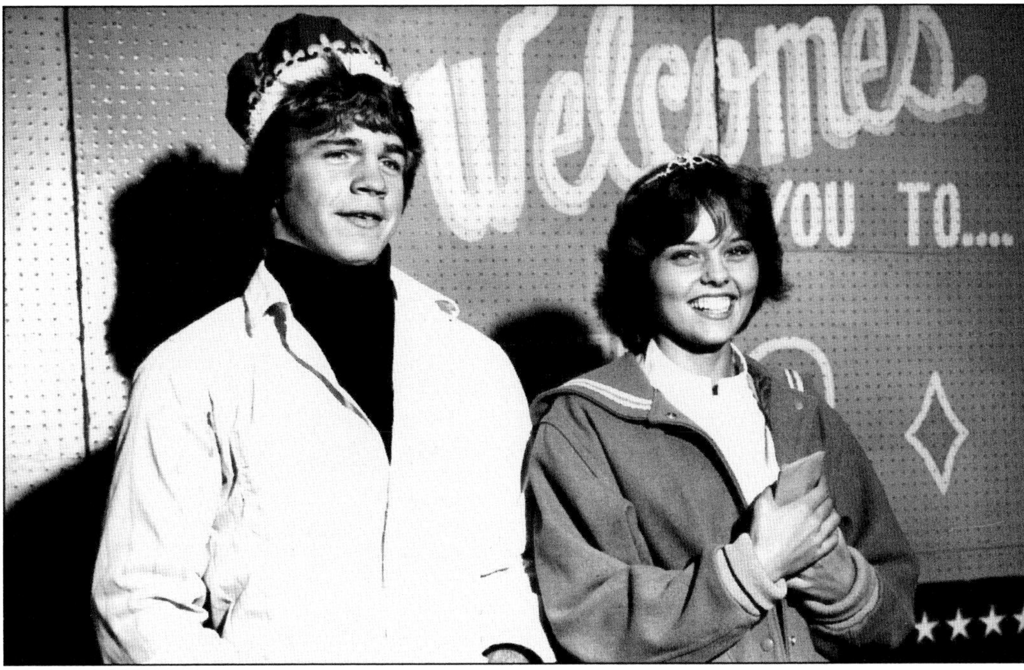

The 1979 fair king, Mike Clark from South Dearborn, and queen Volina Stultz from Switzerland County stand onstage and help the judges hand out ribbons to various prize winners. (J. F. Ullrich collection.)

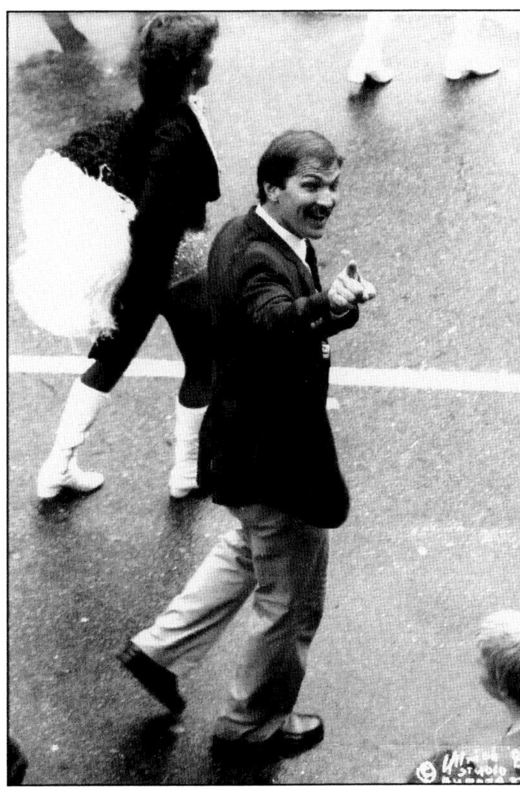

South Dearborn band director David Kling points to the camera as he follows along with the band. (J. F. Ullrich collection.)

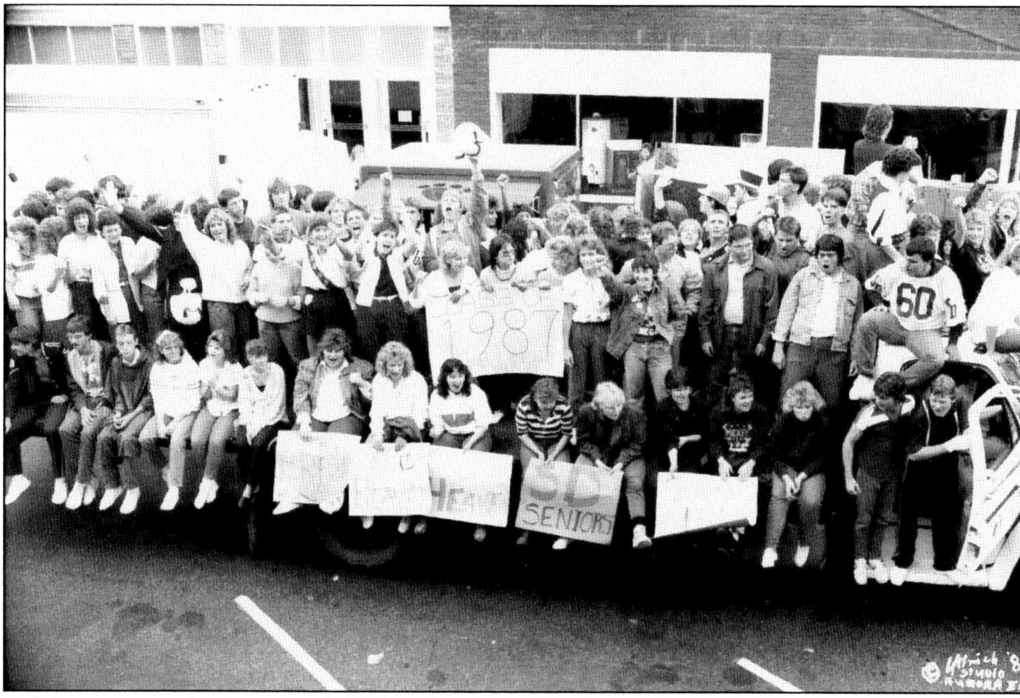

Celebrating its last year of high school is the South Dearborn class of 1987. (J. F. Ullrich collection.)

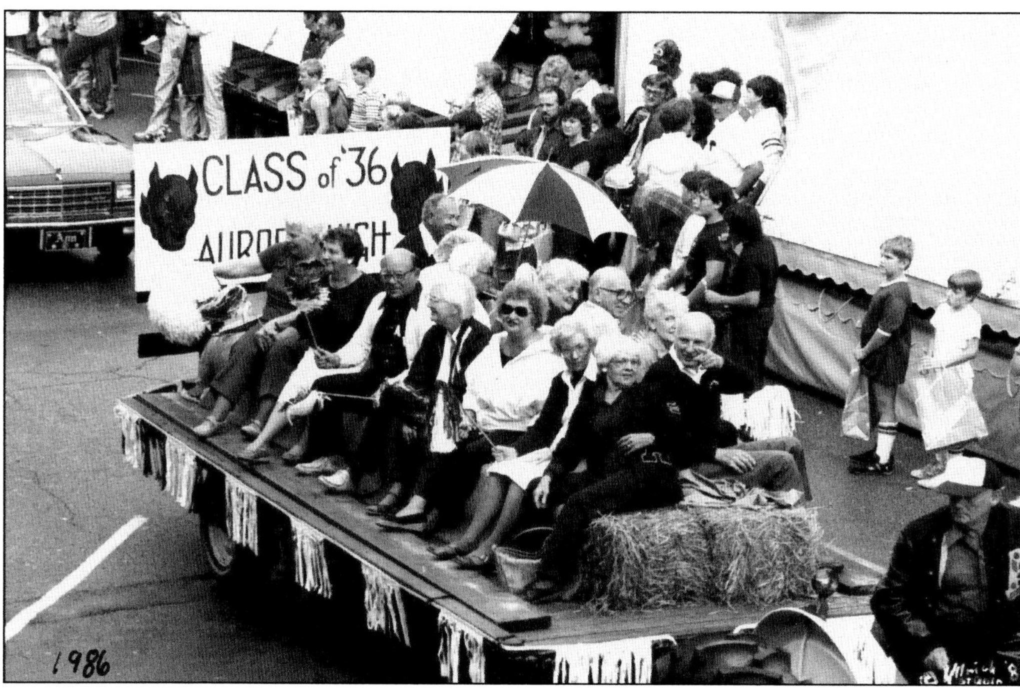

Like many classes, the Aurora High School class of 1936 chose farmers fair weekend to celebrate its 50th class reunion. (J. F. Ullrich collection.)

For many years, Indiana's Ninth District representative Lee Hamilton rode through the parade. He served in the House of Representatives from 1965 to 1999. Known as a master negotiator, he cochaired the Iraq Study Group and the 9-11 Commission. (R. Ullrich collection.)

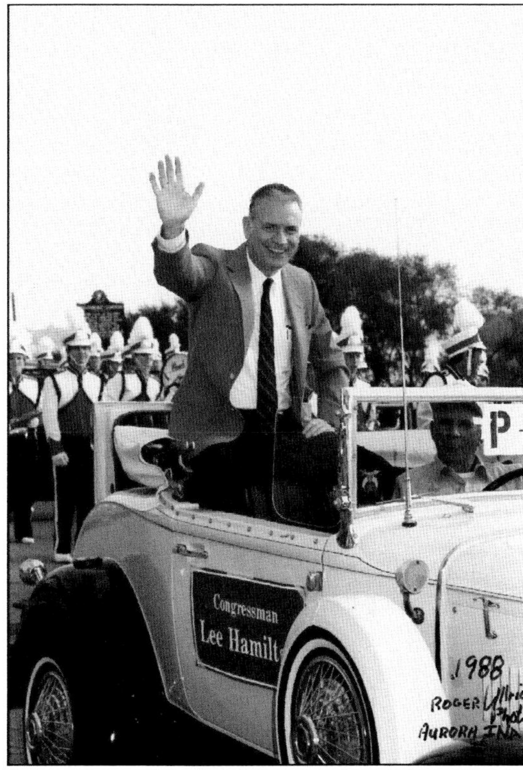

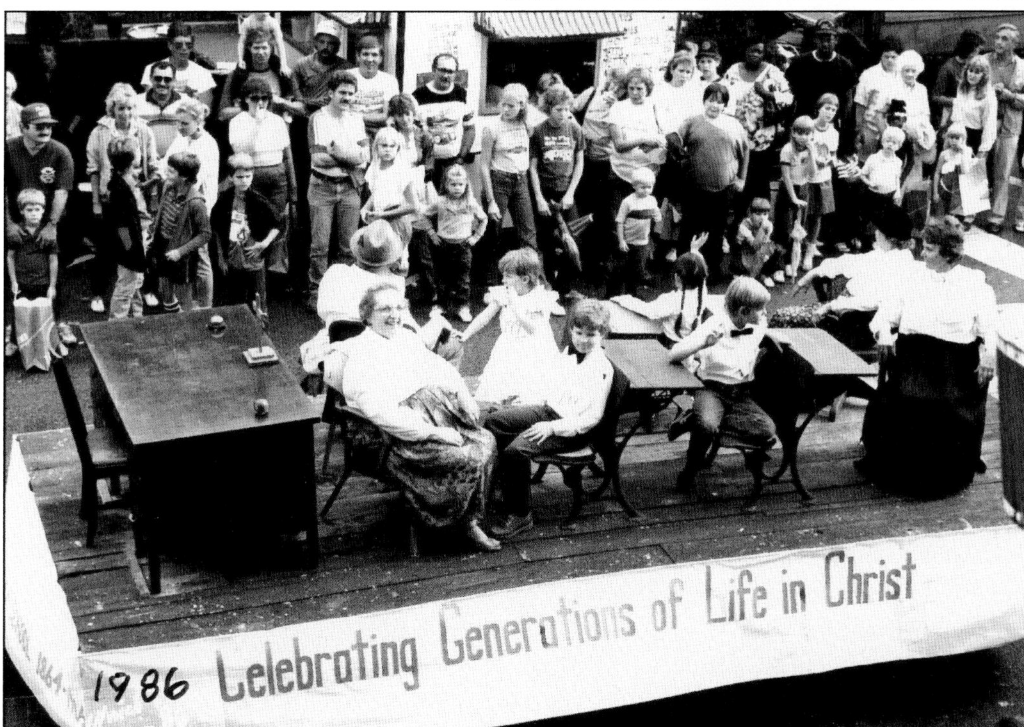

St. John's Lutheran School displays an old one-room schoolroom. (J. F. Ullrich collection.)

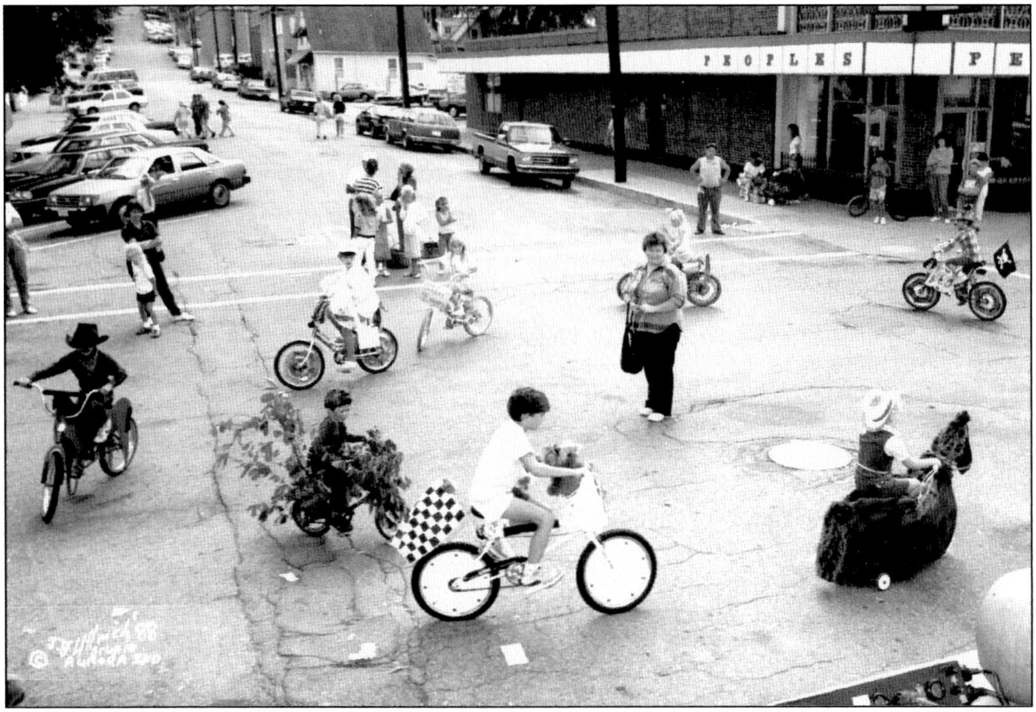

In 1988, a camouflaged soldier, cowboys, and pirates circle in front of the stage on their bicycles for the best decorated. (J. F. Ullrich collection.)

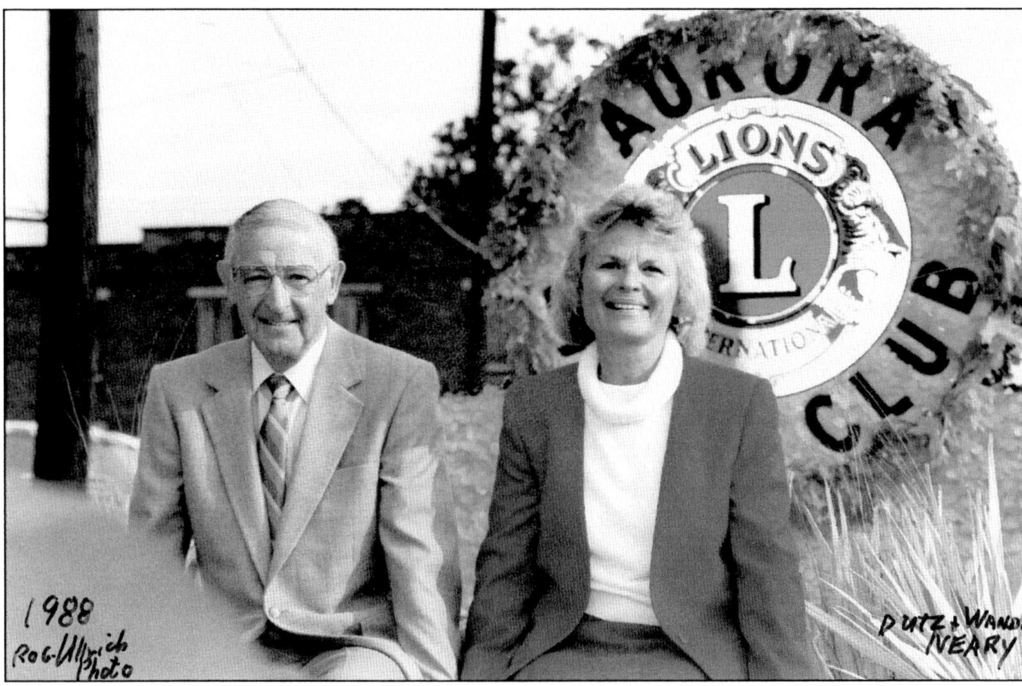

In 1988, the grand marshal was Dr. Sylvester Walter "Deutz" Neary. Seated on the float with him is his wife, Wanda. Deutz was chosen for his longtime commitment to the Lions Club. As an Aurora businessman, he was the founder of Neary and Burton Optometrists.

Aurora mayor Leon Kelly and Lawrenceburg mayor Carl Agner pose together at the 1988 fair. (R. Ullrich collection.)

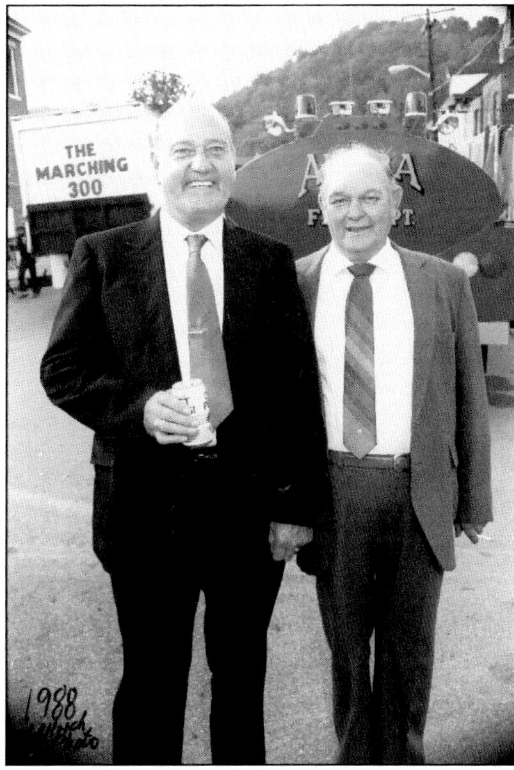

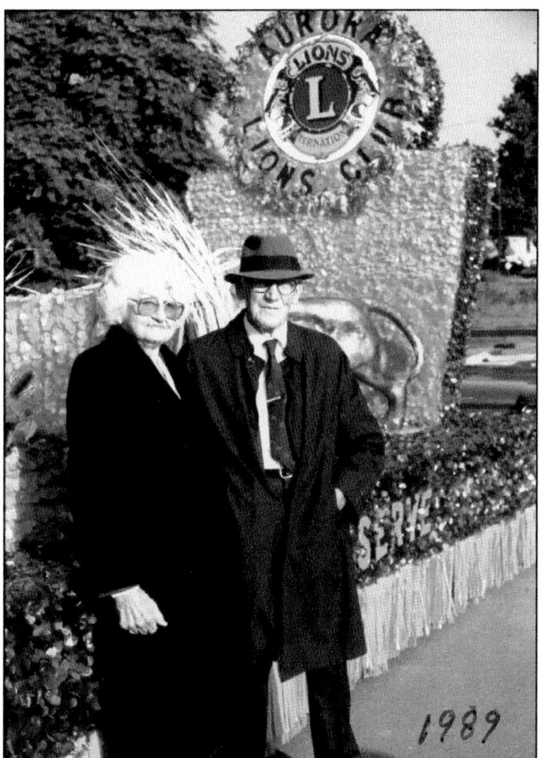

Grand marshals Gene Gabbard and his wife, Ruth, pose in front of their parade float. Between them, they lay claim to 84 years of teaching school. They often touched that "kid in all of us," the theme of the 1989 parade. Following their careers as educators, the Gabbards volunteered and donated much of their time to Aurora. (J. F. Ullrich collection.)

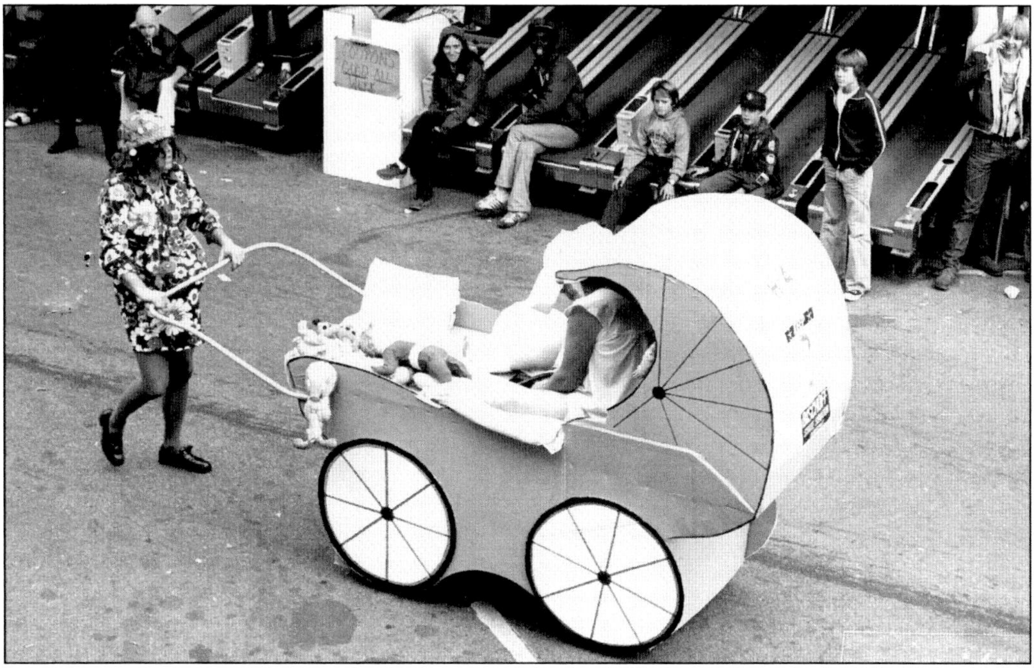

Bob and Frank Linkmeyer combined their efforts to win the most humorous award of all entries in the 1978 fair. Bob, in the buggy, portrays the part of a baby, while Frank played the part of an expecting mother to capture the top award in their division. The buggy is constructed on a riding lawn mower and was self-propelled. Bob nursed a calf's bottle along the parade route while Frank threw diapers to onlookers. (J. F. Ullrich collection.)

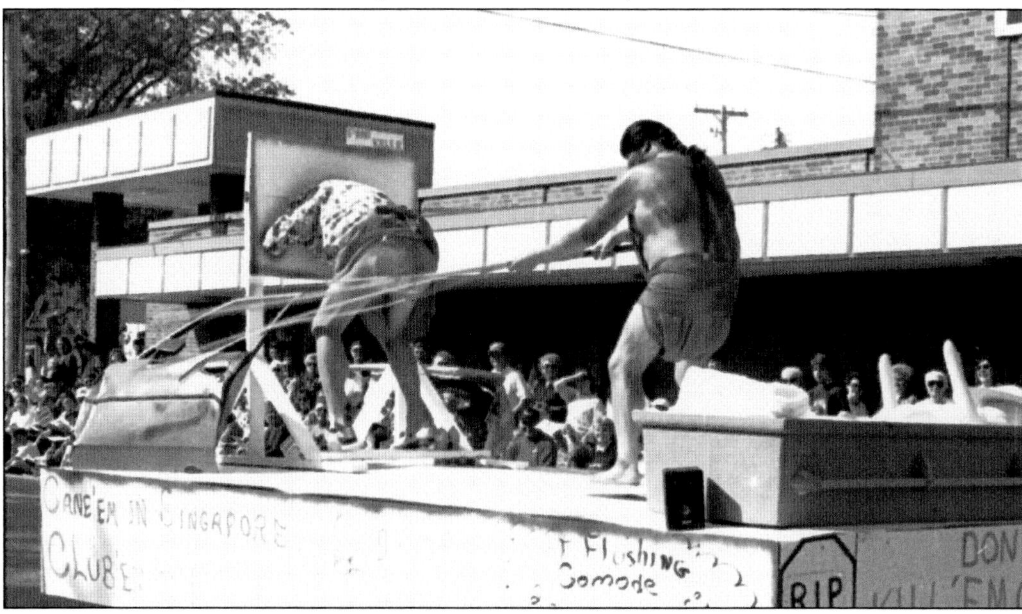

In the 1994 world news, an American boy was caned as punishment in Singapore. Not afraid to make fun of themselves or anyone else, brothers Bob and Frank Linkmeyer demonstrate the Aurora version of the event. Bob swings his cane as a harsh punishment on Frank's "bare bottom." Frank received six swats for not flushing the commode. (Aurora Lions Club collection.)

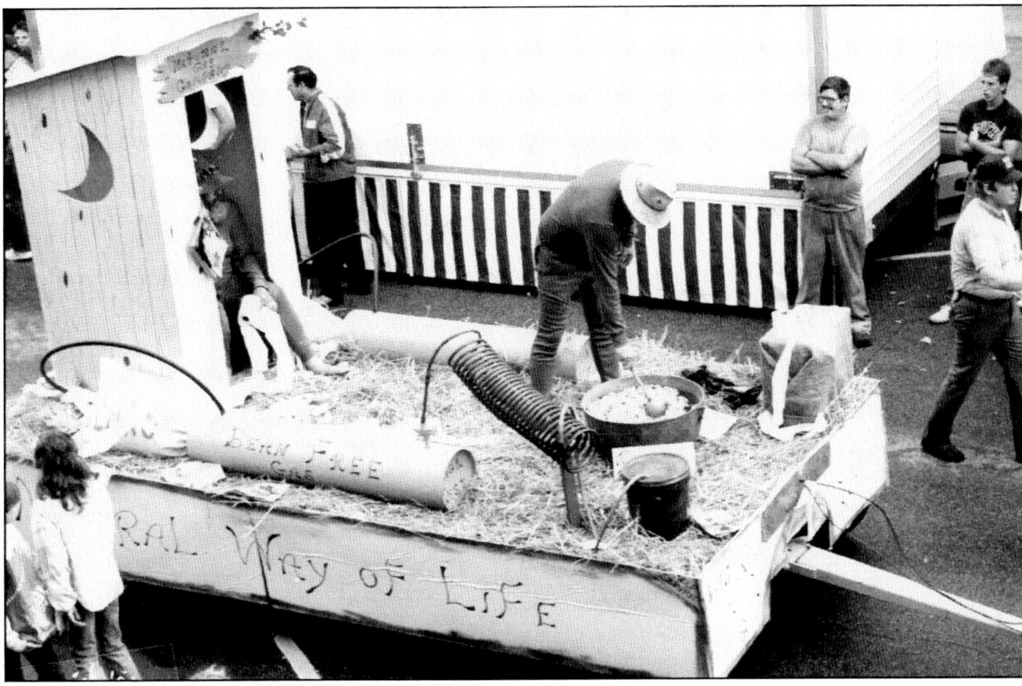

By creating free gas, Bob and Frank Linkmeyer demonstrate a way of conserving our natural resources. One brother stirs a pot of beans heated with gas generated by the other brother in the outhouse. (J. F. Ullrich collection.)

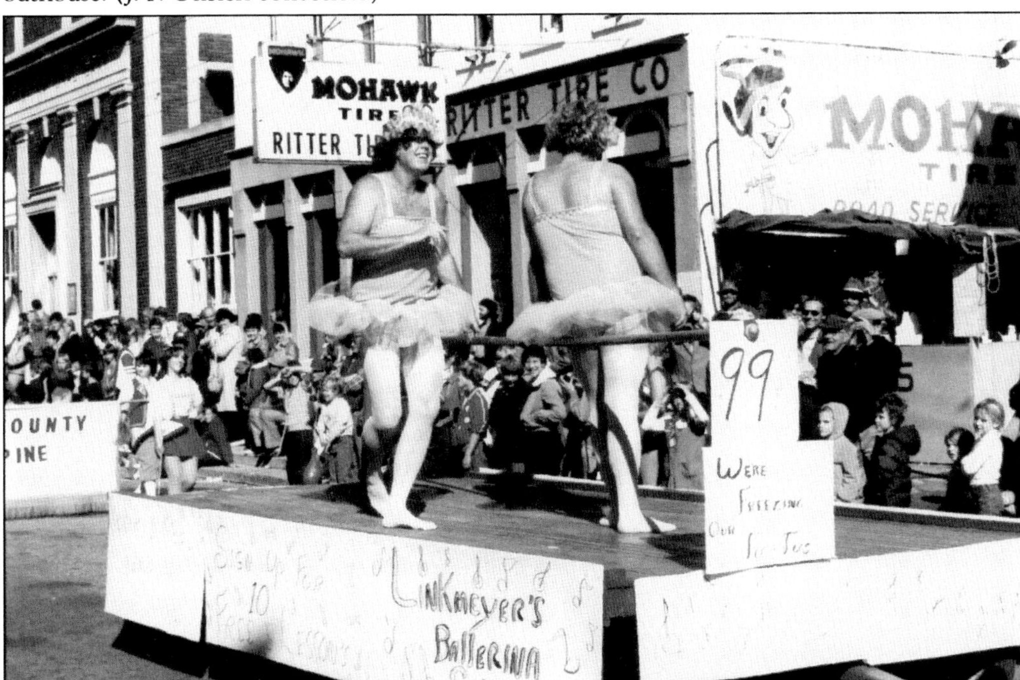

Light on their feet, Bob and Frank Linkmeyer dance for the crowd in oversized tutus. Jokingly they promote the "Linkmeyer Ballerina School." (Aurora Lions Club collection.)

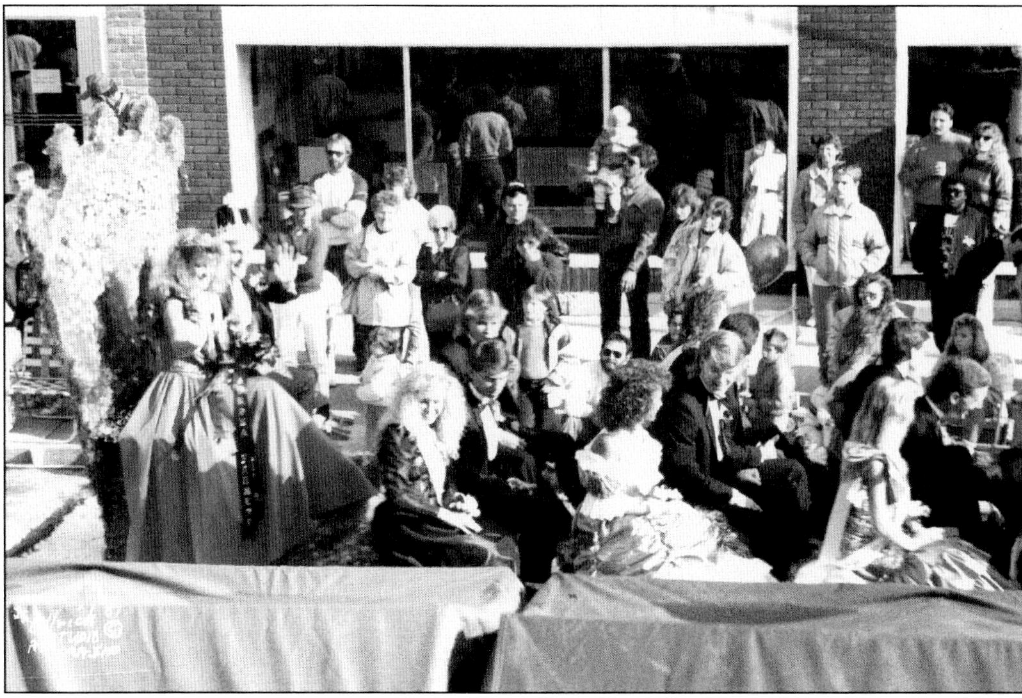

King Geoffrey Heil of Moores Hill and queen Jennifer Leap of the Southeastern Career Center wave to the crowd from their throne. Other candidates could not be identified. (J. F. Ullrich collection.)

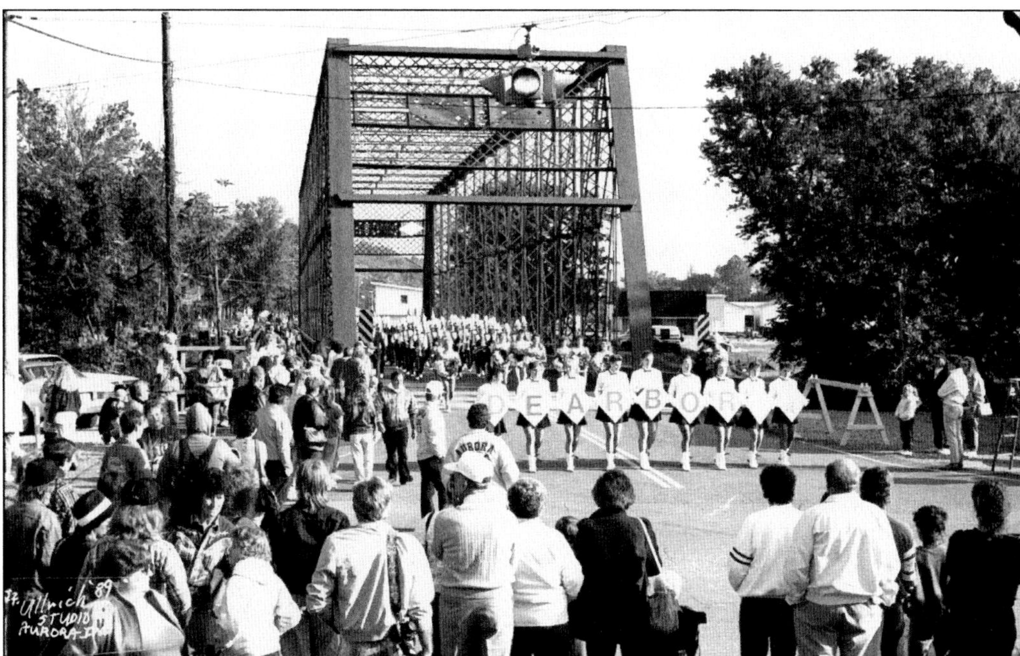

The 1989 South Dearborn Band crosses the George Street Bridge. (J. F. Ullrich collection.)

Five

FROM 1990 TO 2008

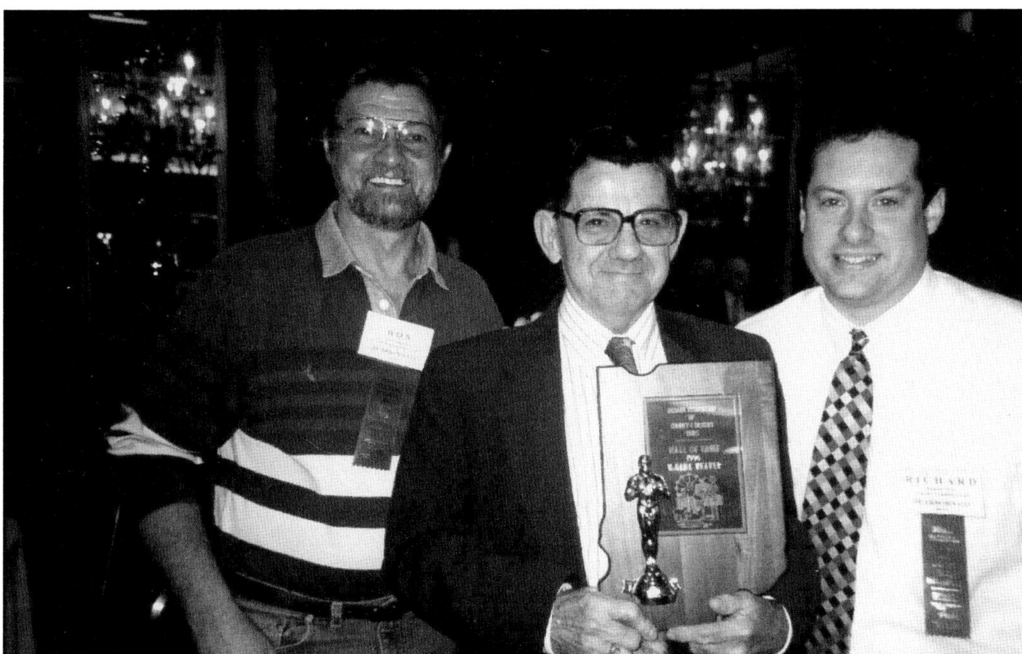

Lions Club member Gene Weaver served as master of ceremonies for the Aurora Farmers Fair for nearly 32 years. On Sunday, January 5, 1997, Weaver (center) was the recipient of the hall of fame award from the Indiana Association of County and District Fairs. Pictured with him are Aurora Farmers Fair board members Ron Goodpaster, secretary (left), and Richard Ullrich, vice president (right). (Richard Ullrich collection.)

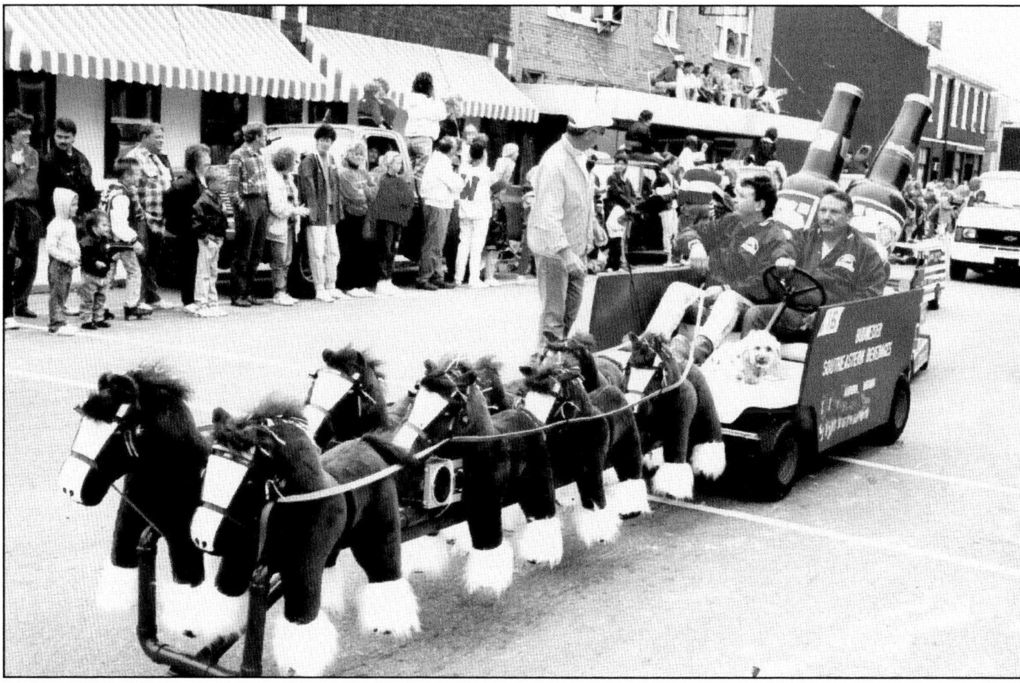

A team of Clydesdale horses pull the Southeastern Beverage wagon. The Budweiser distributor is located on Industrial Drive, Aurora, and owned by Robert Hastings. (Aurora Lions Club collection.)

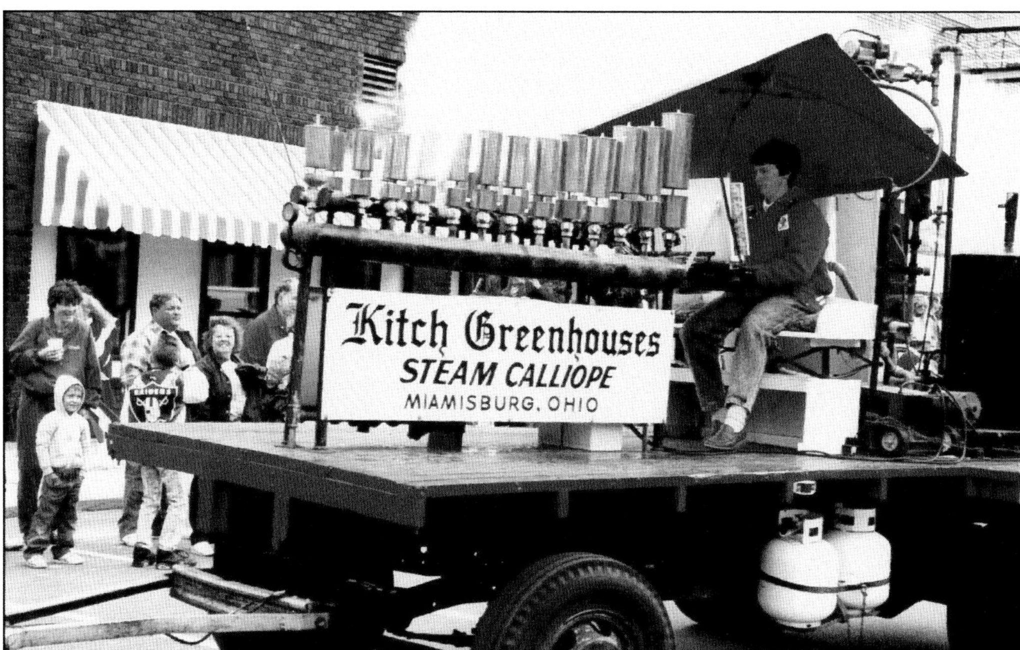

Kitch Greenhouses steam calliope from Miamisburg, Ohio, plays a tune for the crowd. (Aurora Lions Club collection.)

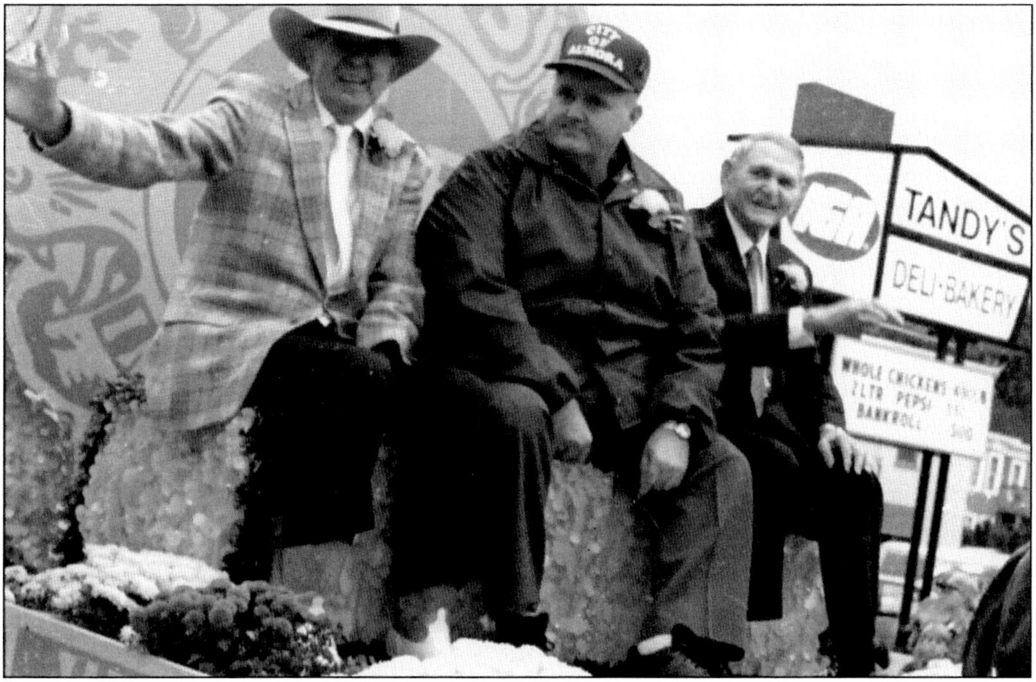

In 1991, three men were chosen for the honors of grand marshal: Raymond Baer (left), Wally Slayback (center), and Harry Kyle (right). For many years Baer played Santa Claus for many children. Slayback took great pride in his work for the city. Kyle also gave much of his time to the betterment of Aurora. (Aurora Lions Club collection.)

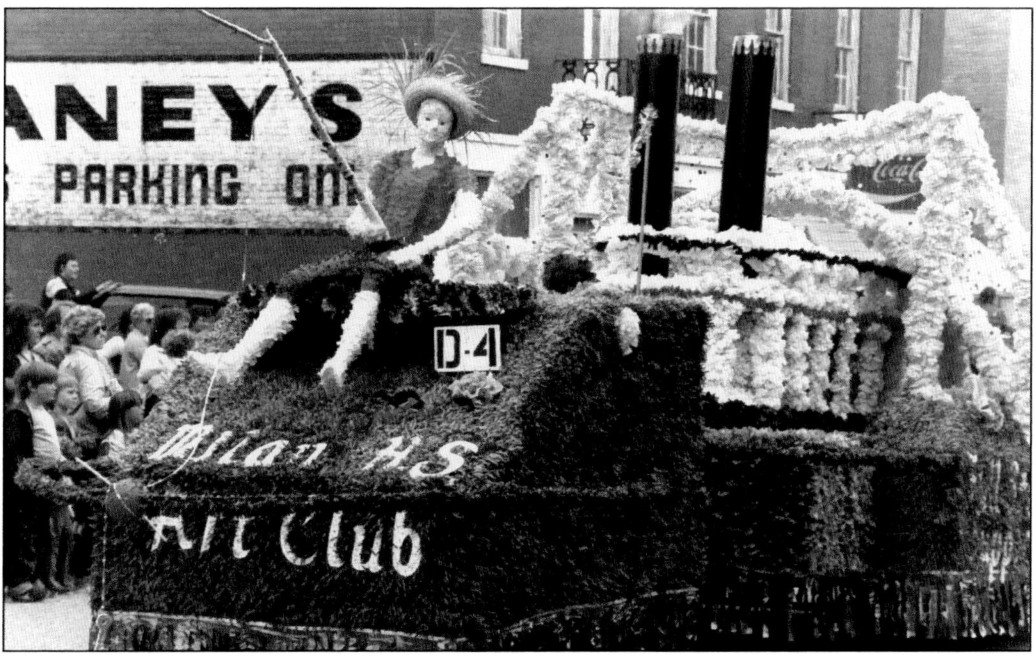

Milan High School Art Club built this elaborate float. The young man fishes off the front of a contained vehicle. A riverboat sails under a bridge on the back part of the float. (Aurora Lions Club collection.)

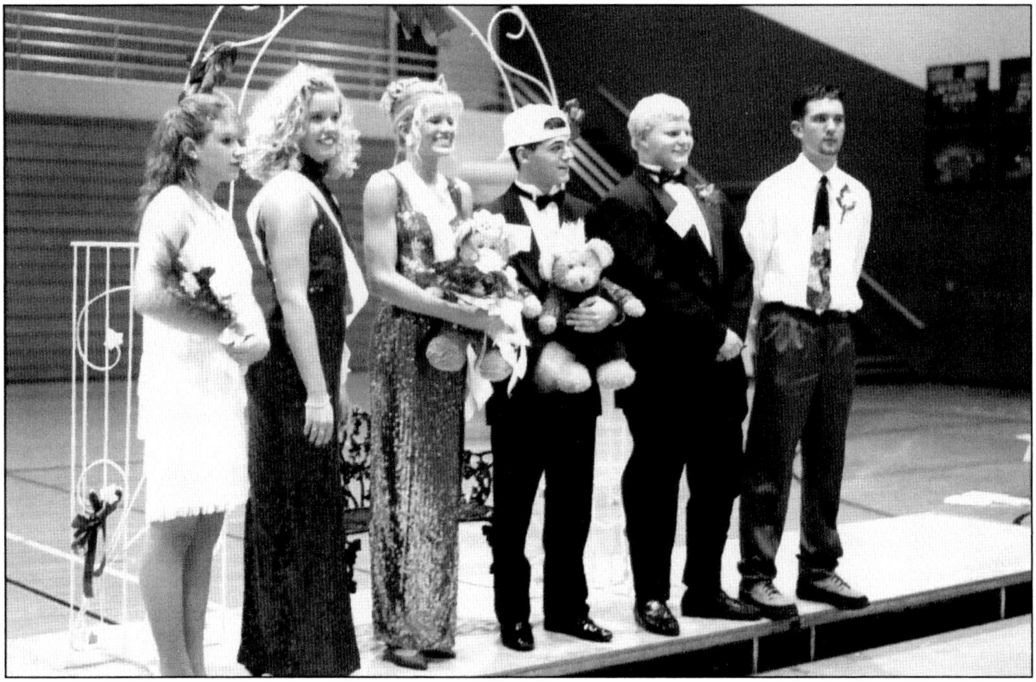

Rain did not dampen the spirit of this 1995 king and queen competition. The program was moved to the South Dearborn High School. Reigning queen Christa Dahmann crowned Dianne Buxton from Jac-Cen-Del High School queen and Philip Saragoza as king from South Dearborn High School.

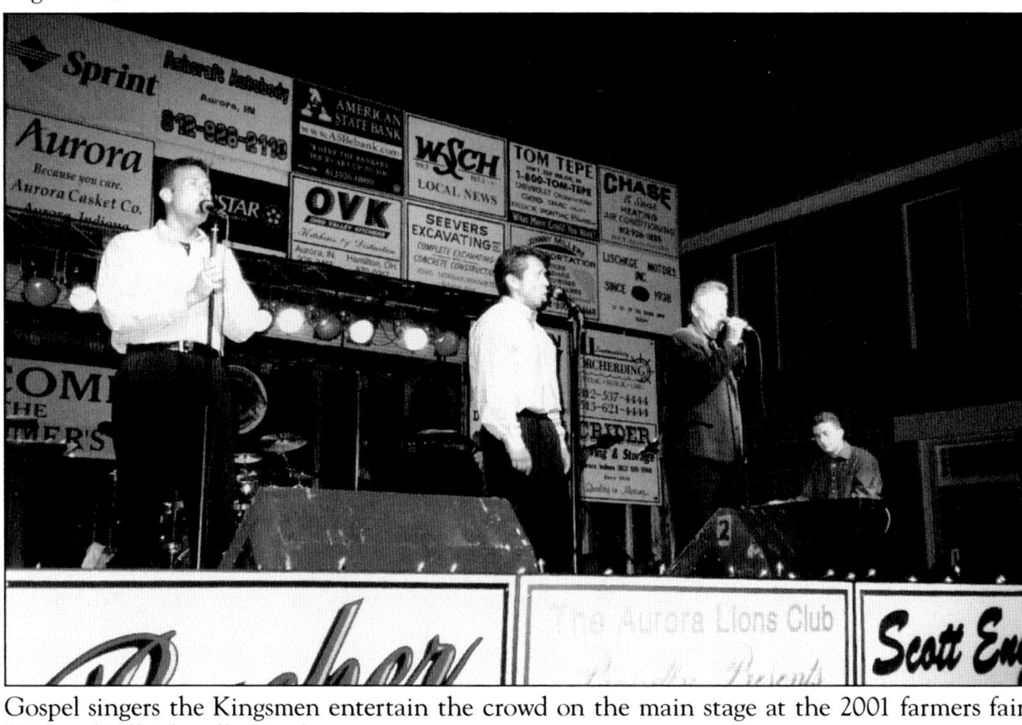

Gospel singers the Kingsmen entertain the crowd on the main stage at the 2001 farmers fair. (Richard Ullrich collection.)

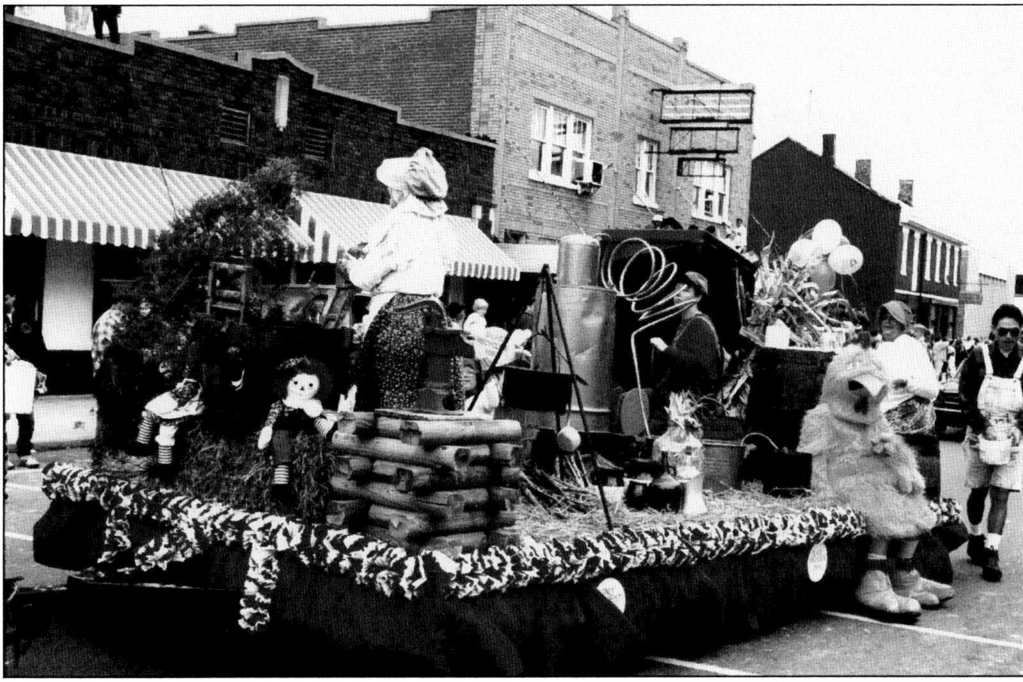

A farmer and his wife, along with their pet chicken, make their way up Third Street. (Aurora Lions Club collection.)

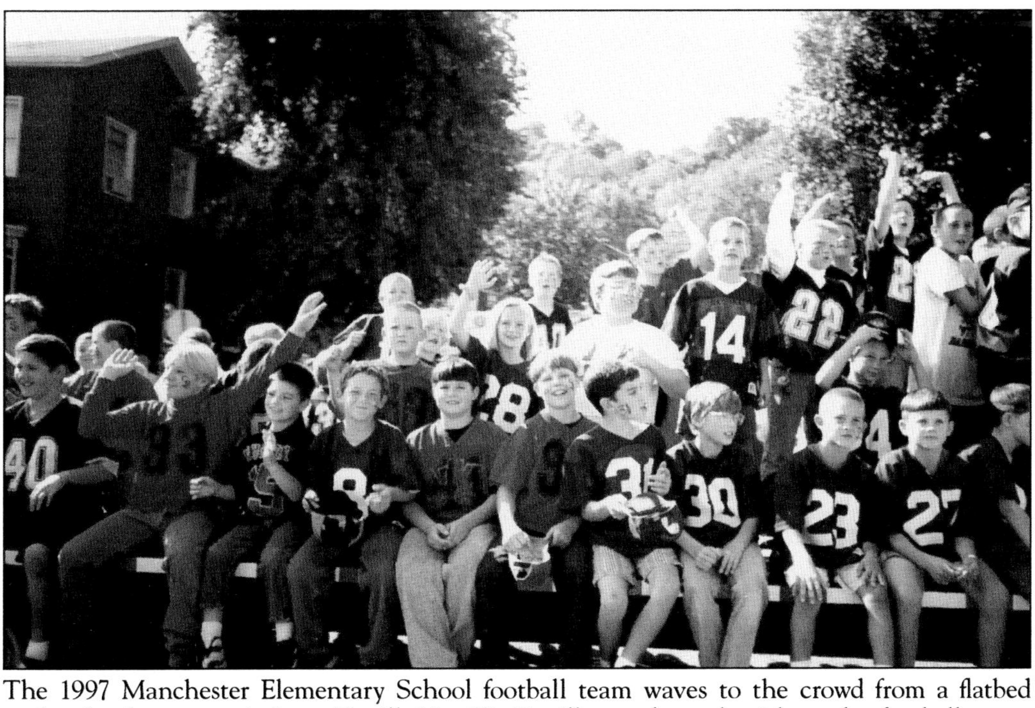

The 1997 Manchester Elementary School football team waves to the crowd from a flatbed trailer. In the center is Jesse Terrill, No. 28. Terrill was the only girl on the football team. (Largent collection.)

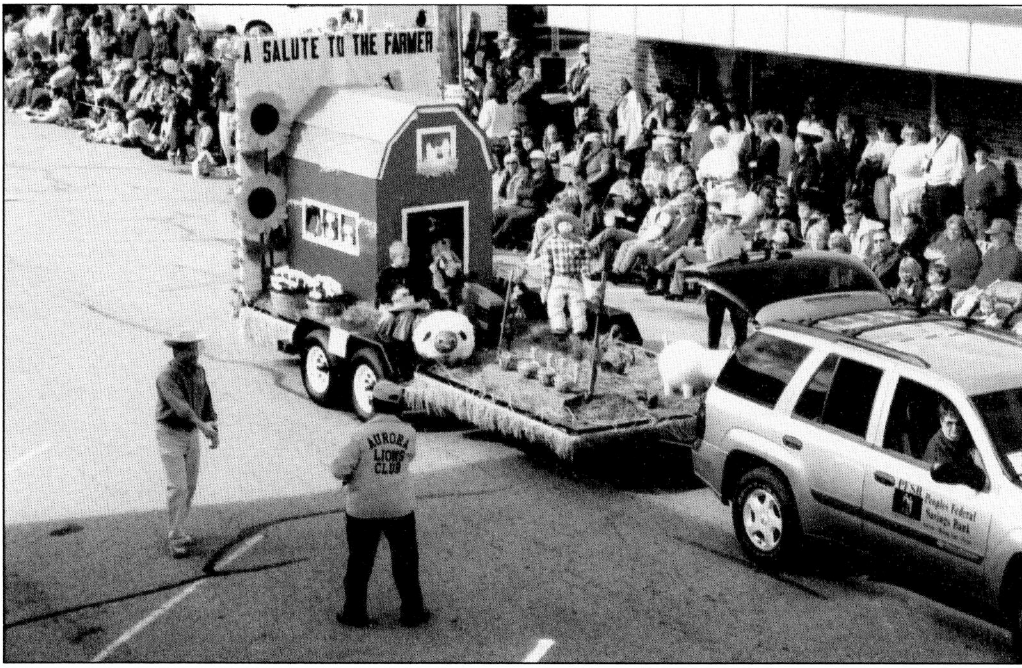

Peoples Federal Savings Bank pulls a barnyard float named A Salute to the Farmer. (Largent collection.)

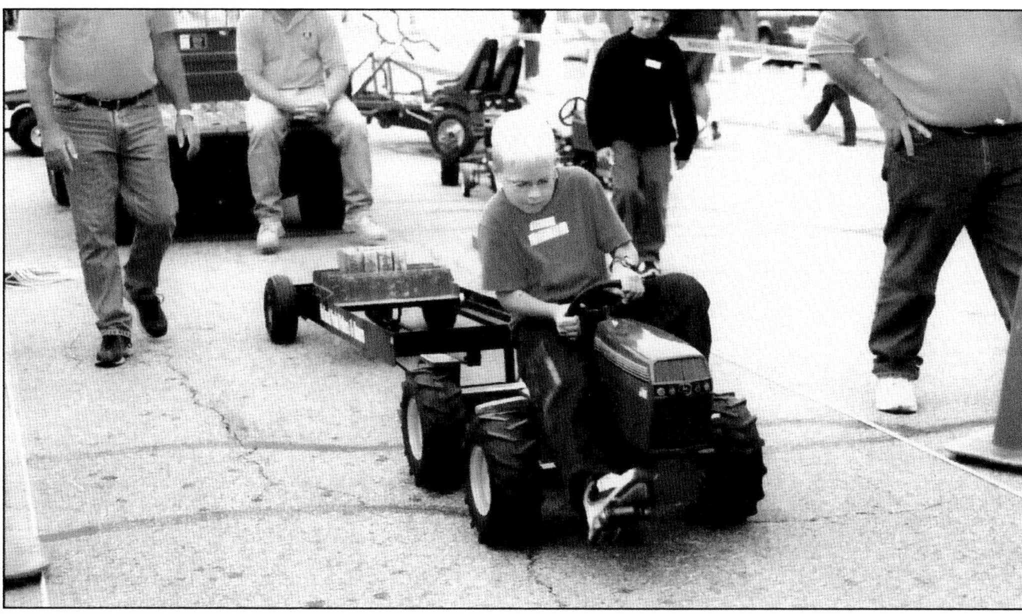

Tommy Jackson won first place at the kids tractor pull on Bridgeway Street. Young Jackson pulled the sleigh 28 feet and 9 inches, winning $10 and ride tickets for his efforts. (Largent collection.)

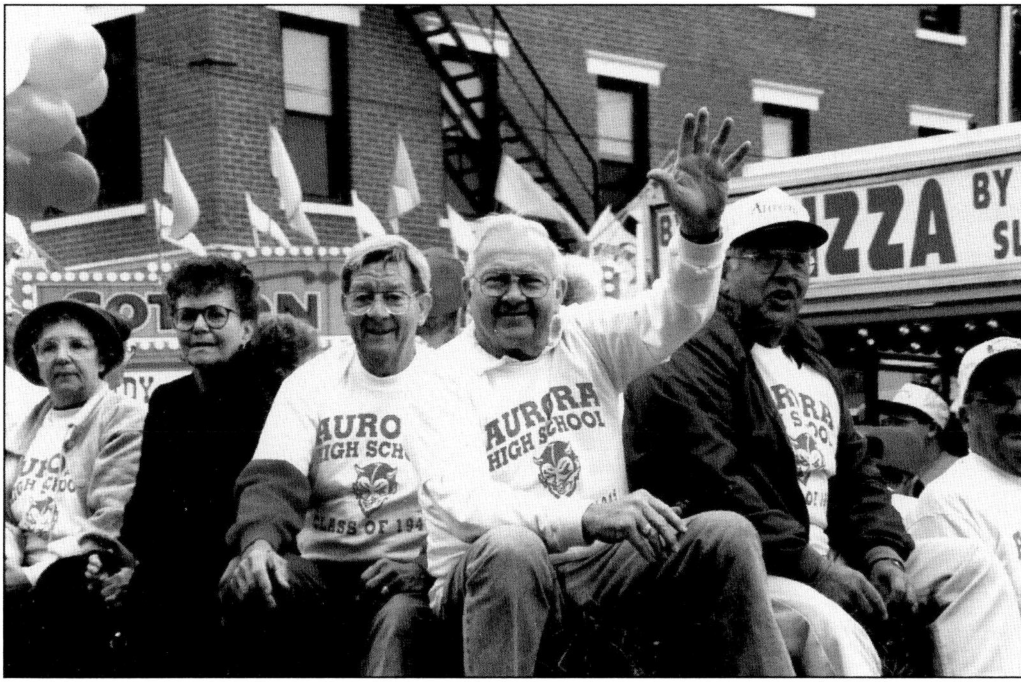

Alumni represent the Aurora High School class of 1945. Bobby Lisckge gives a big wave to the crowd. Robert "Chick" Kittle is seated on his right. (Aylor collection.)

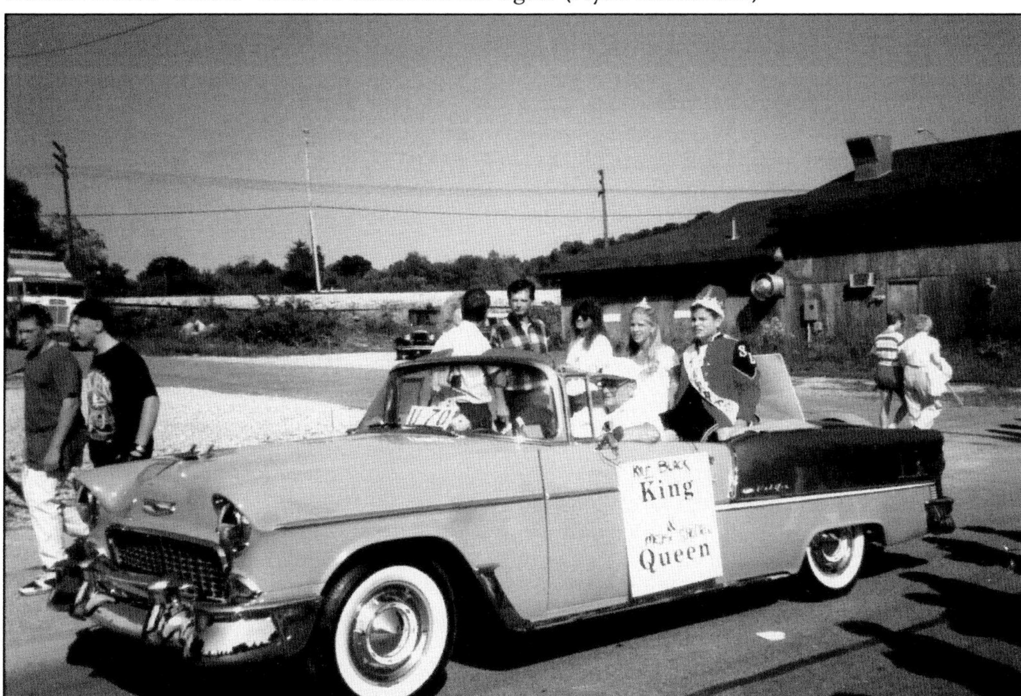

The 1994 Dillsboro homecoming king, Kyle Black, and queen Megan Shelton represent the local festival. They are waiting in the staging area on George Street in a vintage car loaned to them by J. D. Largent. (Largent collection.)

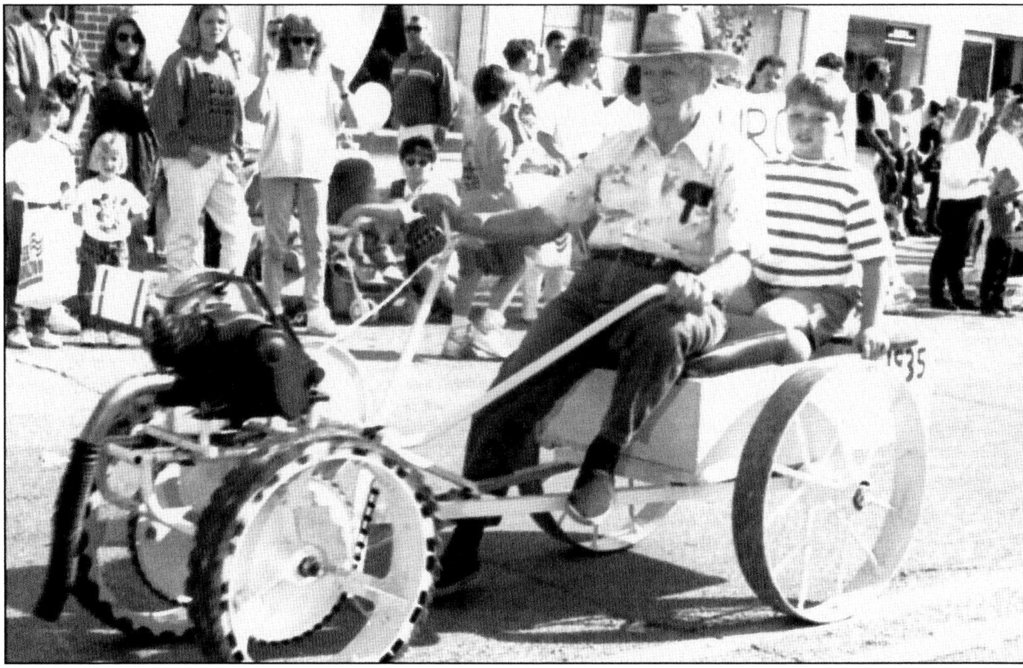

A gentleman and his passenger enter a front-wheel-drive homemade tractor in the parade.

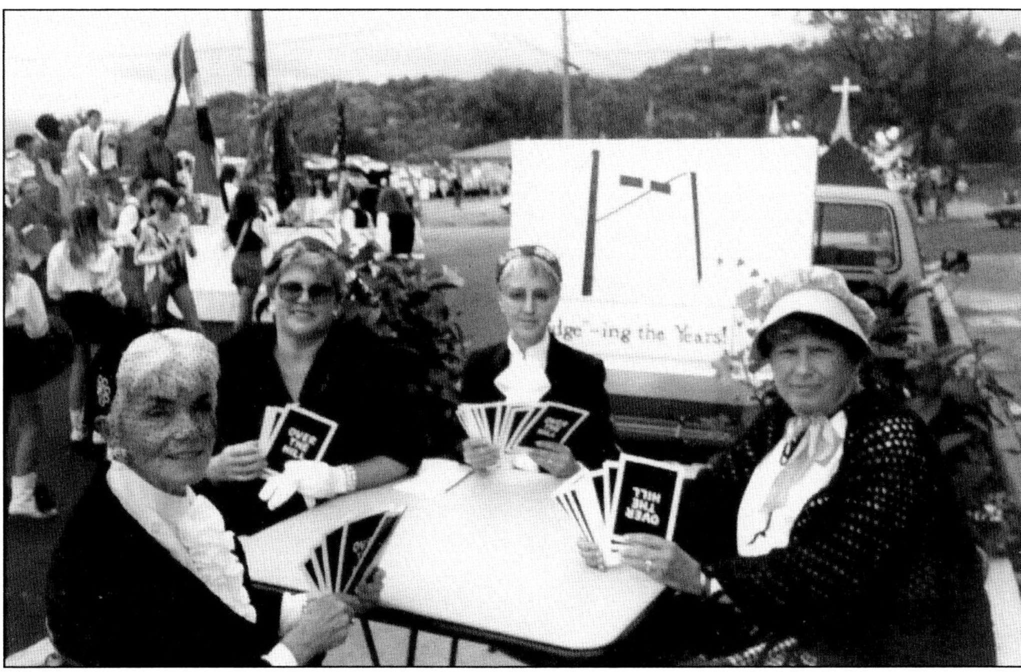

Aurora Women's Bridge Club members Marge Waldon, Jodell Stegner, Cyndi Strzynski, and Phee Ellinghausen played cards throughout the parade in vintage clothing at the 1995 fair. (Waldon collection.)

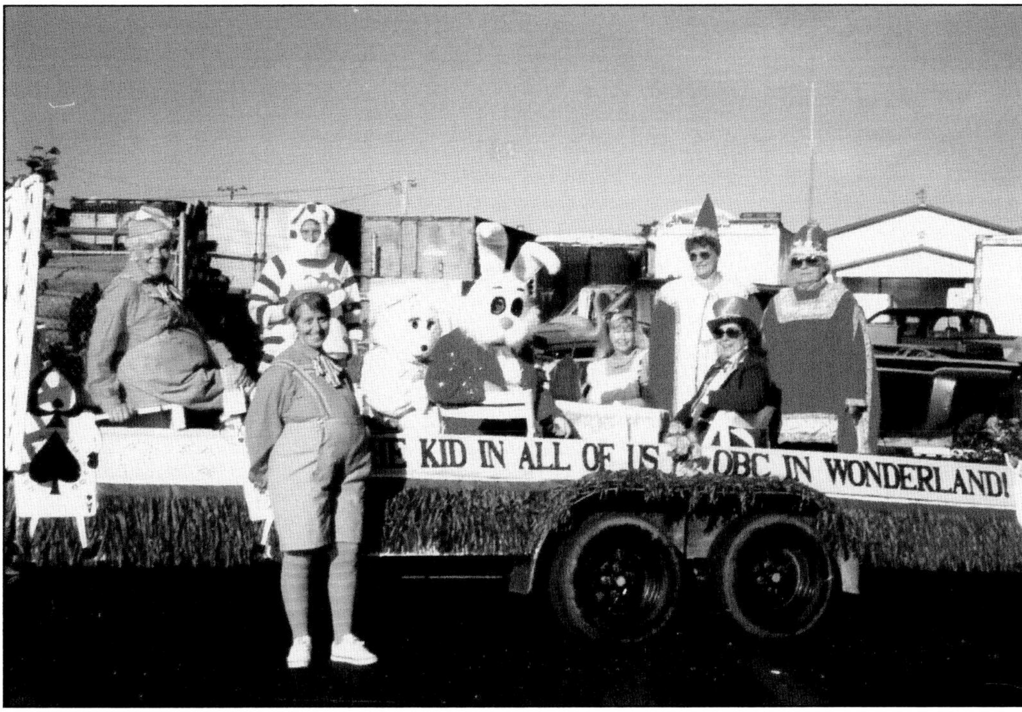

The Kid in All of Us was the theme of the Aurora Bridge Club float in 1996. Not in order are Marge Waldon, Phee Ellinghausen, Cyndi Strzynski, Janet Burton, Charlotte DeVille, Shirley Meyer, and Laura Rolf.

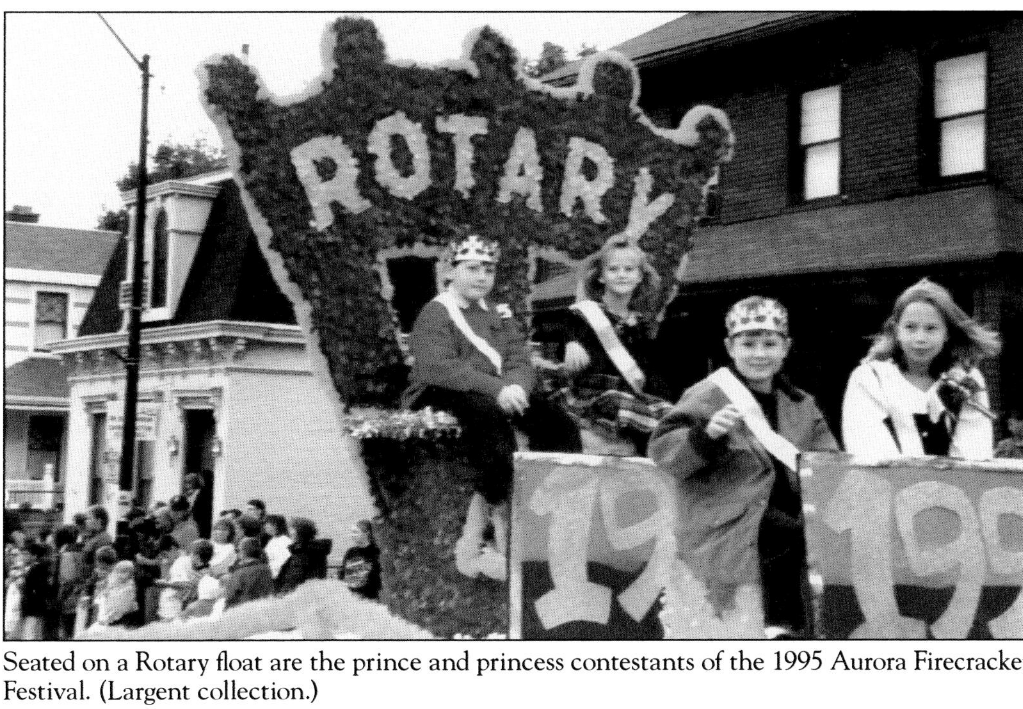

Seated on a Rotary float are the prince and princess contestants of the 1995 Aurora Firecracker Festival. (Largent collection.)

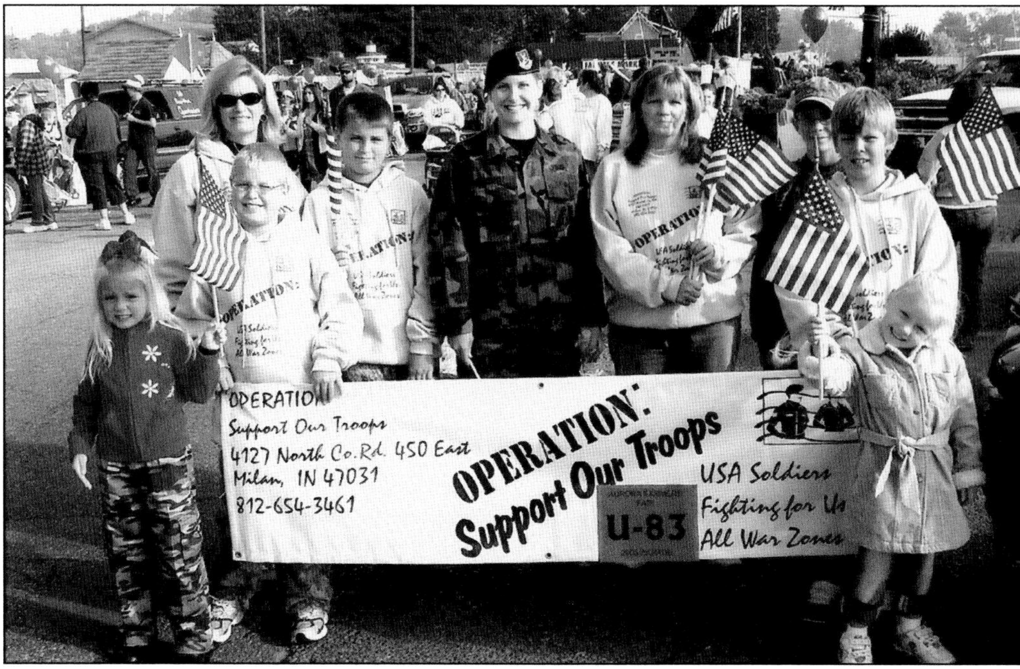

Promoting Operation: Support Our Troops, a group of Milan residents show their support by making up boxes of collected items and sending them on to American soldiers in war zones. From left to right are Kelsie Jackson; Tommy Jackson; Patty Terrill (in back of Tommy Jackson); Trey Jackson; Hannah Terrill, U.S. Air Force; Julie Robinson; Tyler Clark; Matthew Inegart; and Holly Roberts. (Largent collection.)

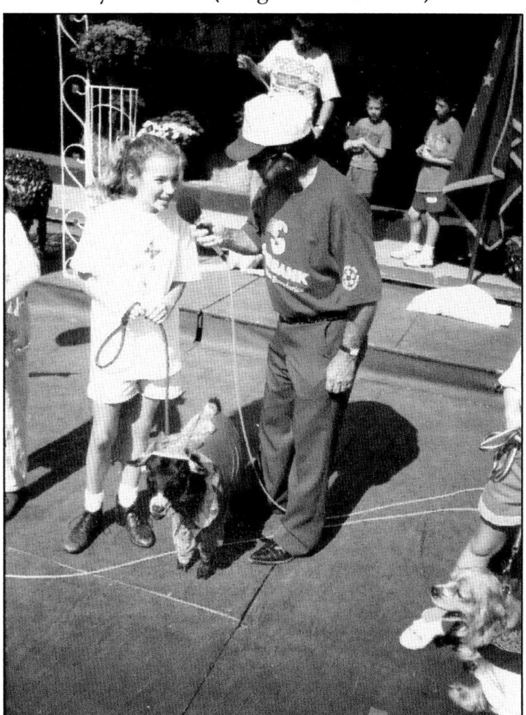

Gene Weaver interviews a young girl and her pet goat at the Friday afternoon pet parade.

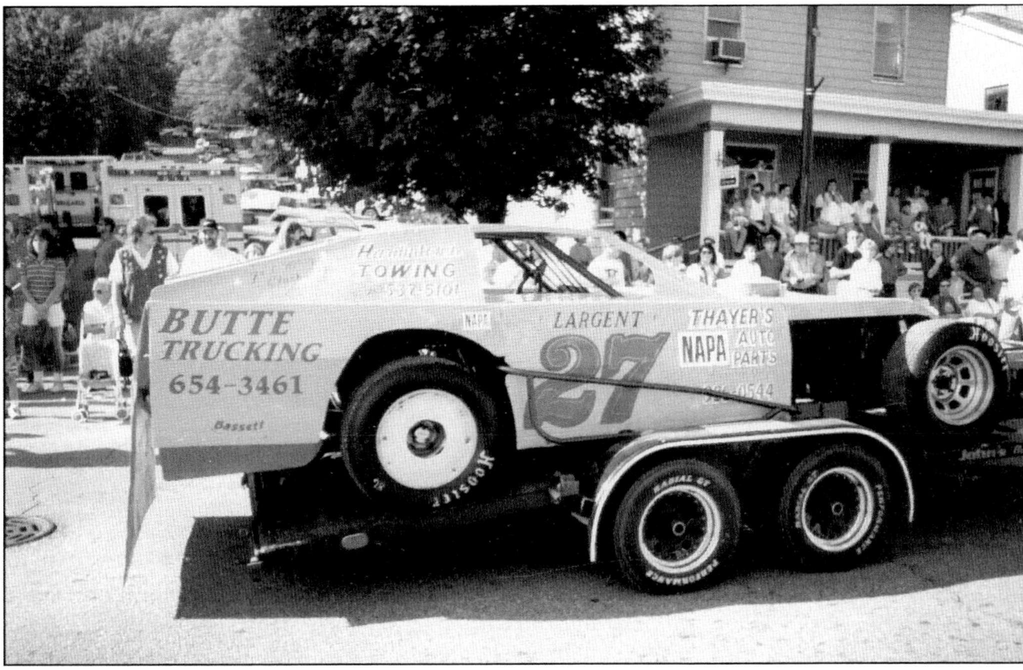

Proud of his win, John D. "Jake" Largent pulls his winning race car throughout the October 4, 1997, parade. (Largent collection.)

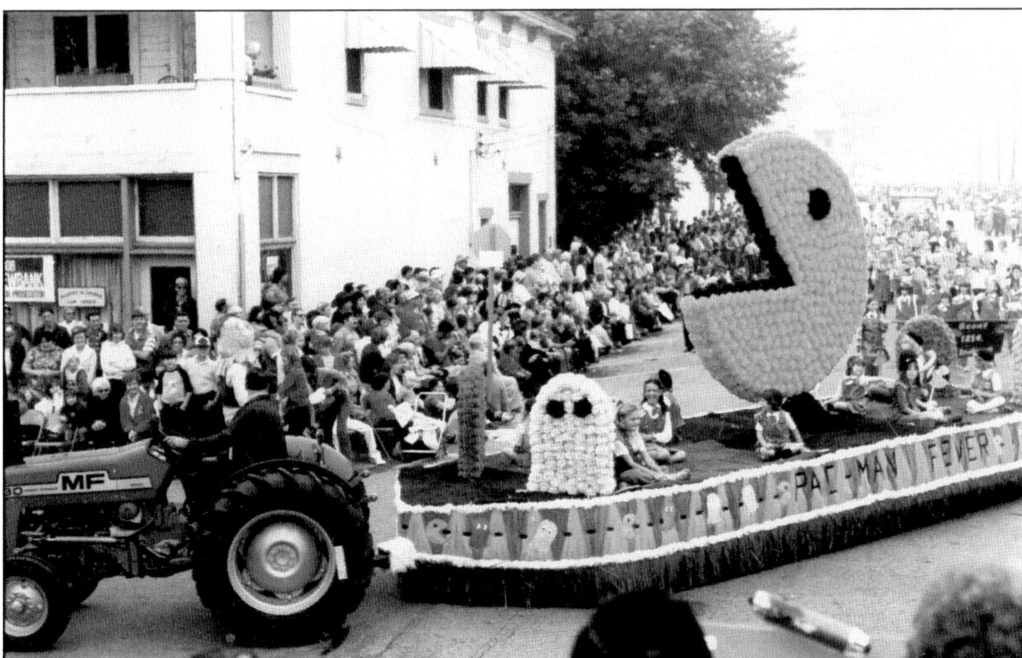

Making its way in front of the stage, the Brownie and Girl Scout troops ride on their Pac-Man float. (Tandy collection.)

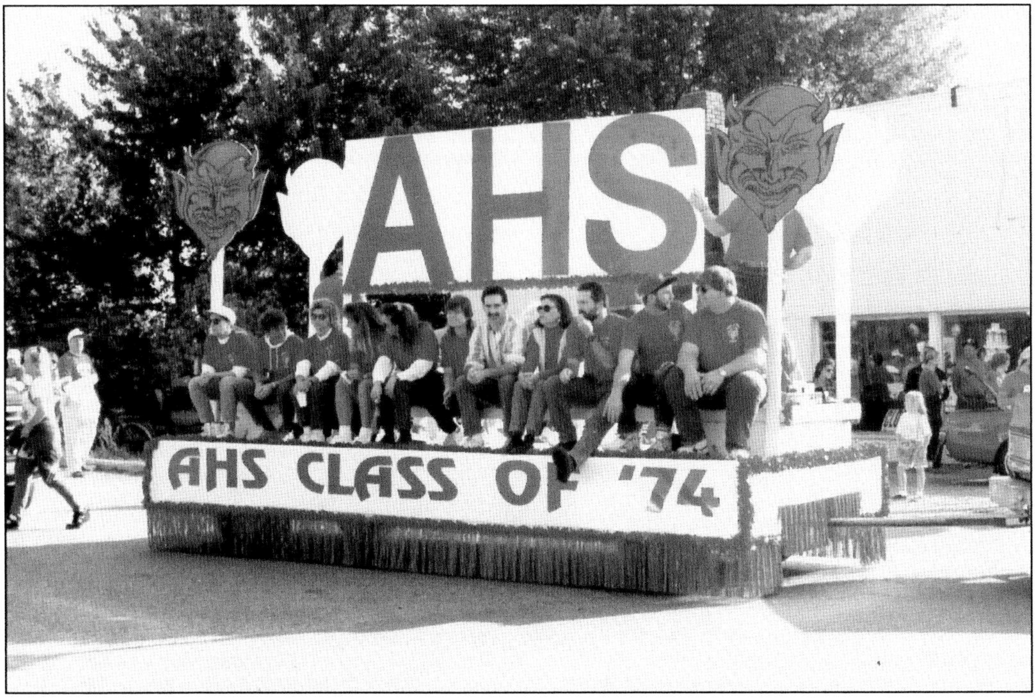

The 1974 Aurora High School class float is seen here. (Richard Ullrich collection.)

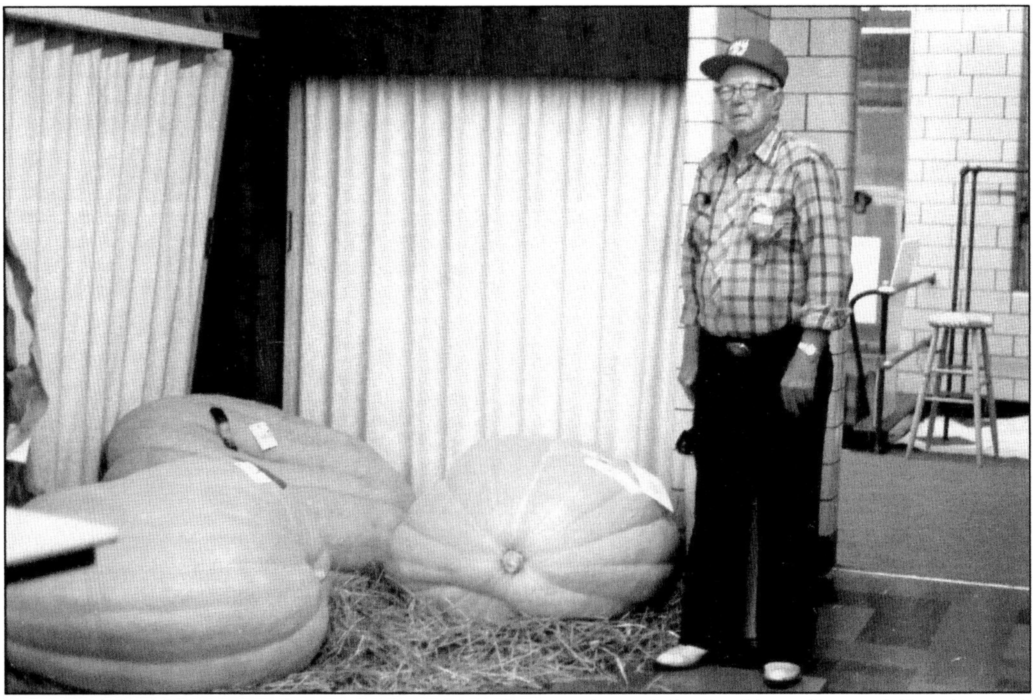

Raymond "Pop" Baer stands next to giant pumpkins in the exhibit building. Retired, Pop donated his time in the exhibit building by keeping an eye on the various displays. He also played Santa Claus for the city for well over 30 years. (Richard Ullrich collection.)

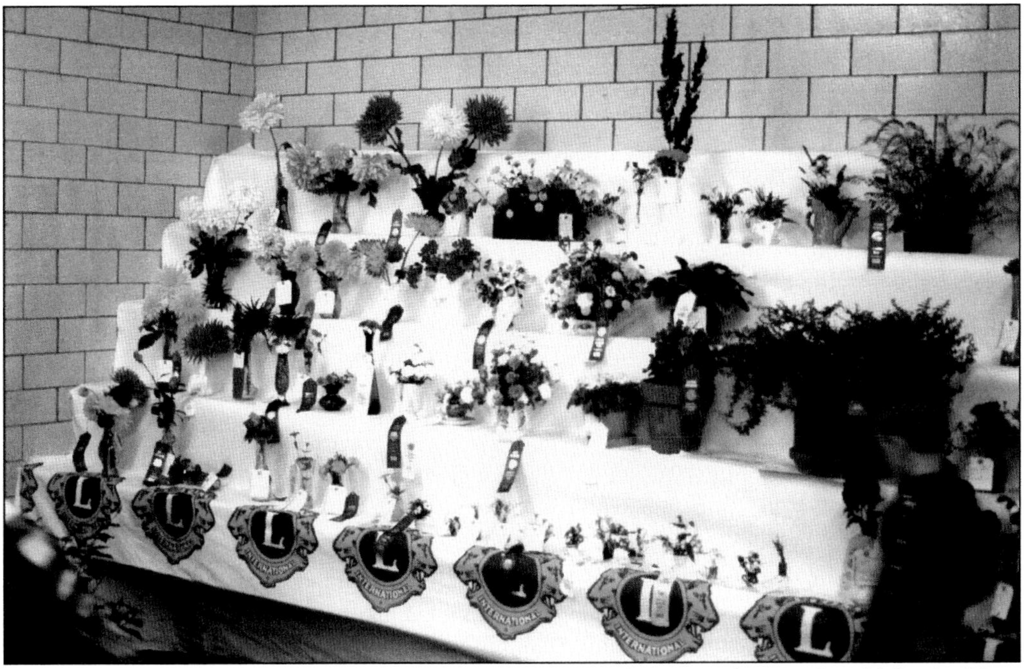

Recently judged flowers stand on display. These photographs were taken in the old fair building on the corner of Main and Second Streets. (Richard Ullrich collection.)

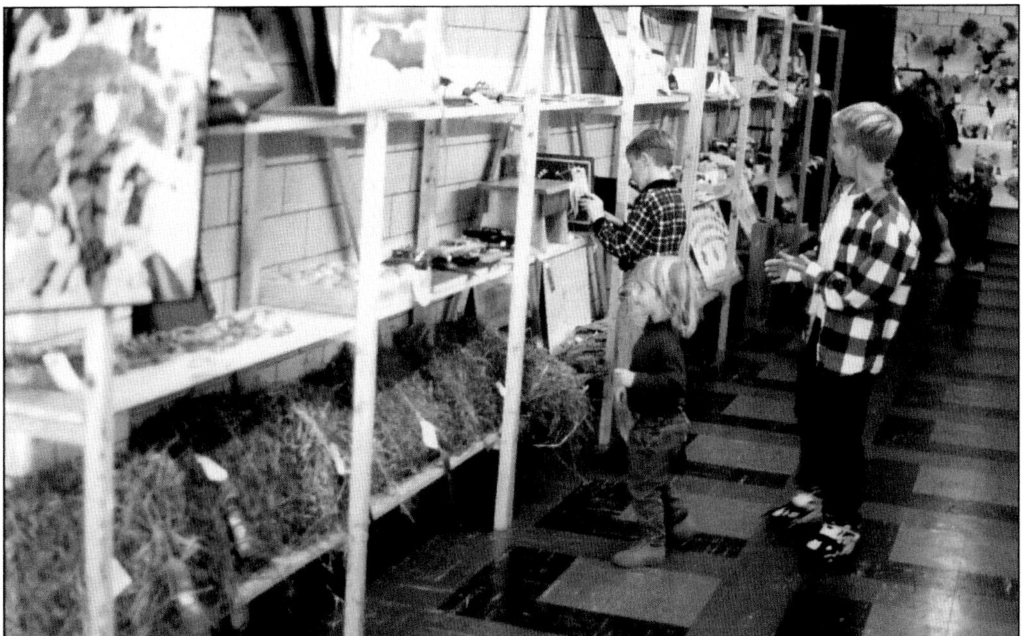

Young children check the various displays located inside the fair building. Young people are encouraged by their schools to enter a project for the fair. (Richard Ullrich collection.)

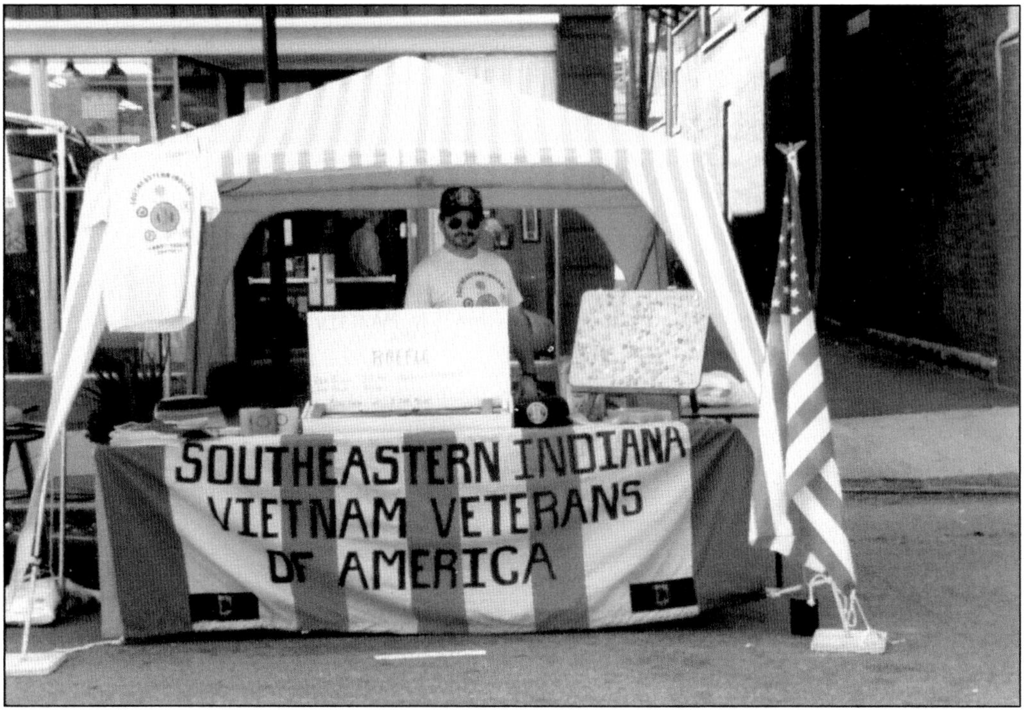

Many booths line the streets advertising and selling raffle tickets. This Southeastern Indiana Vietnam Veterans booth sold raffle tickets and donated the money to their fellow veterans in need.

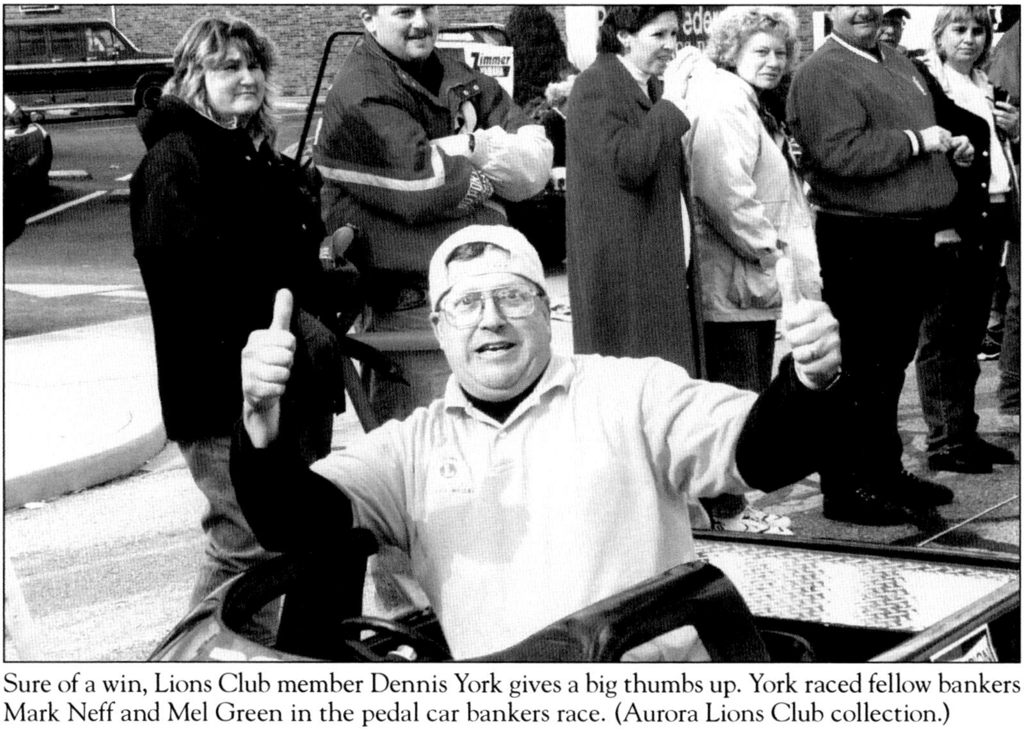

Sure of a win, Lions Club member Dennis York gives a big thumbs up. York raced fellow bankers Mark Neff and Mel Green in the pedal car bankers race. (Aurora Lions Club collection.)

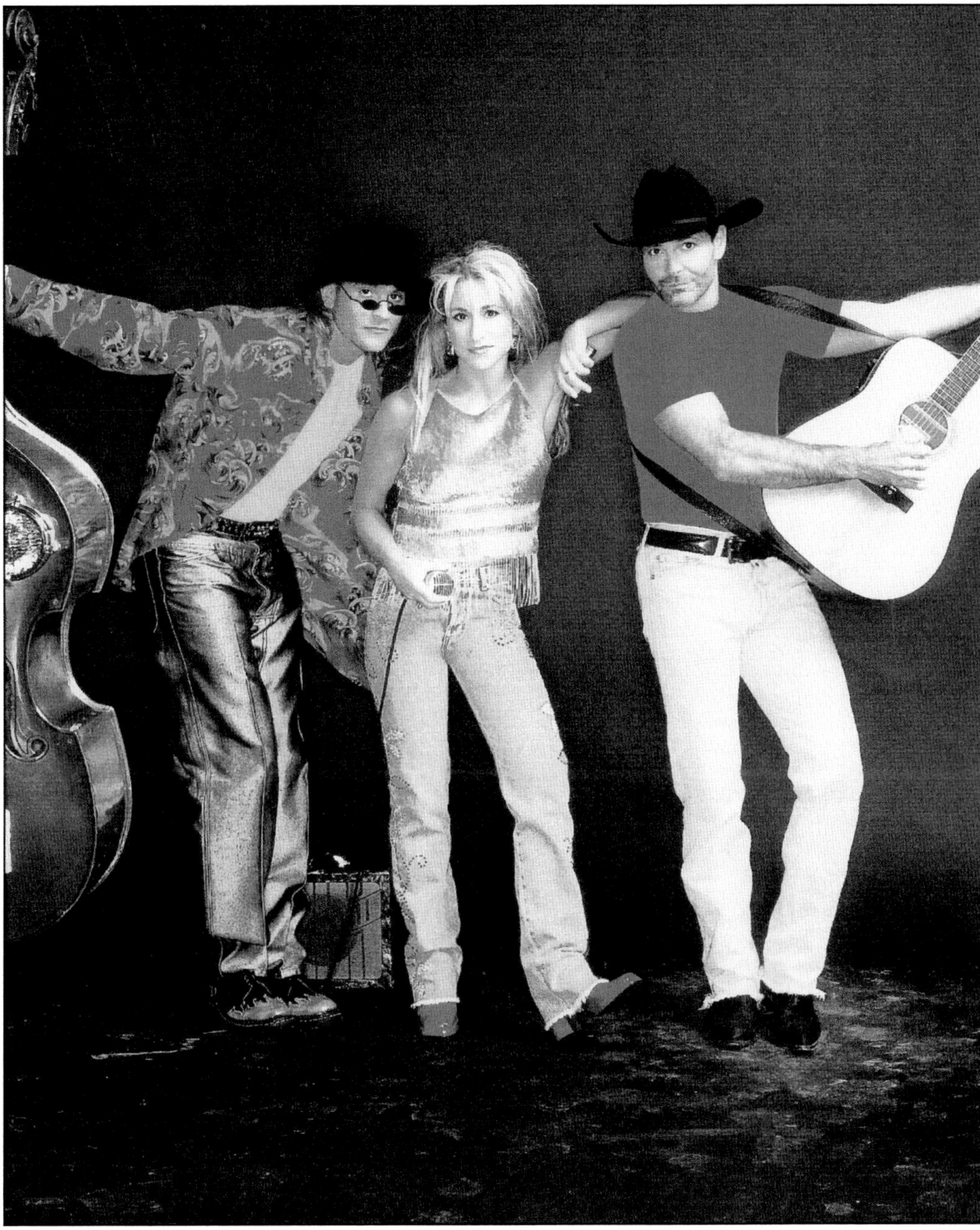

The 2004 Saturday night free entertainment was the contemporary country trio Trick Pony. From left to right, Ira Dean, Heidi Newfeld, and Keith Burns made up the group. The group drew a huge crowd on the Aurora streets as they sang "Pour Me" and other favorite top 10 songs. (Trick Pony collection.)

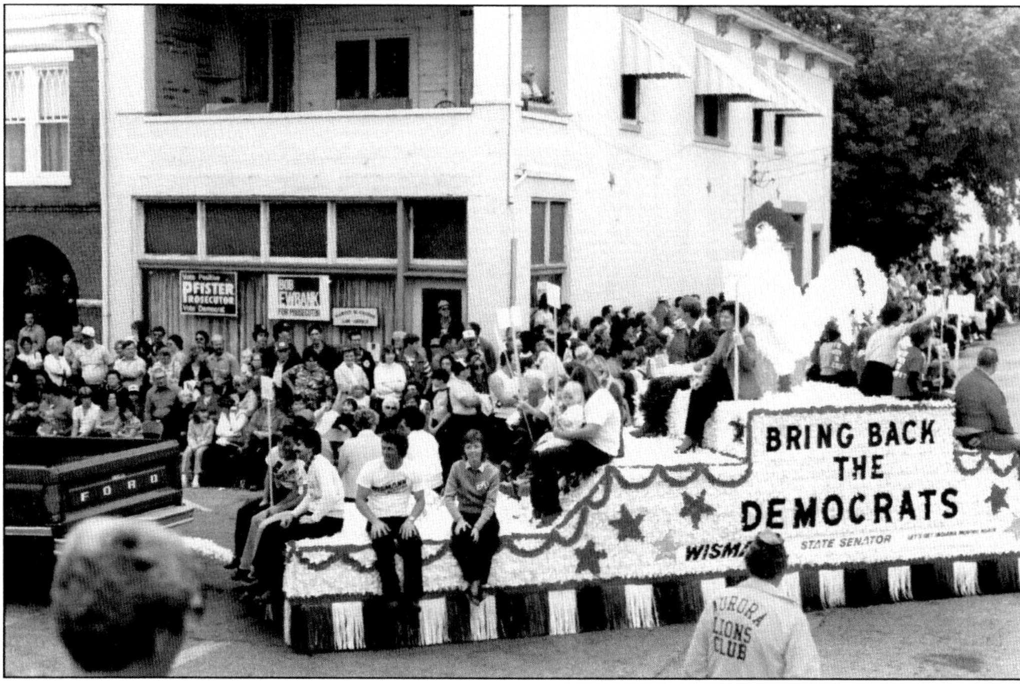

With an election at hand, the Democratic candidates and their families fill the float. (Tandy collection.)

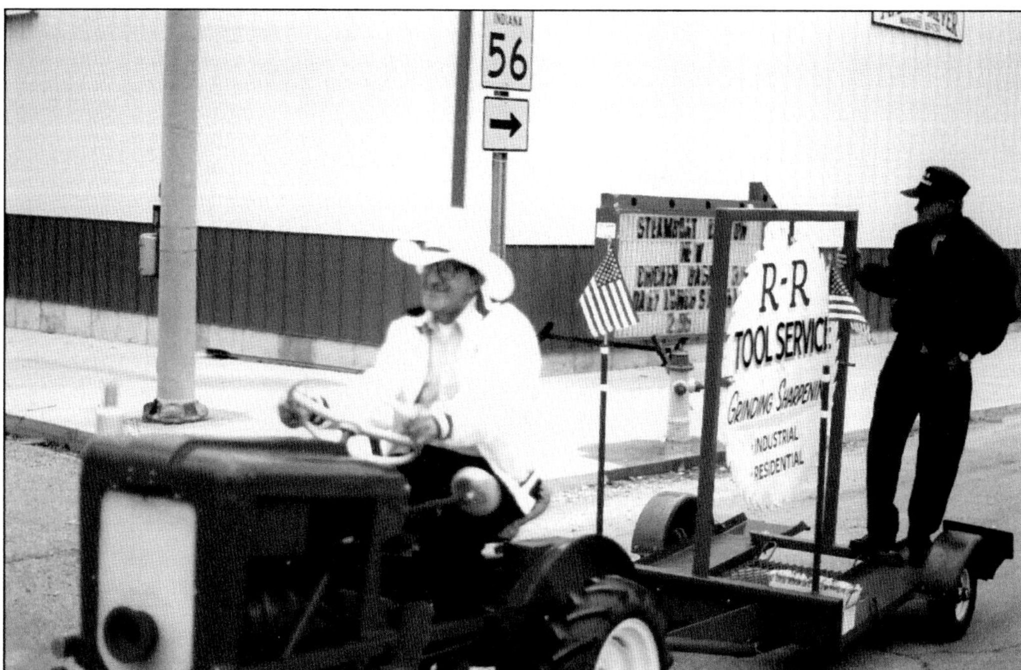

After the parade, Houston Smith is driving the R&R Tool Service display back to the parade staging area on George Street. Catching a ride on back is Robert Nogler. (Richard Ullrich collection.)

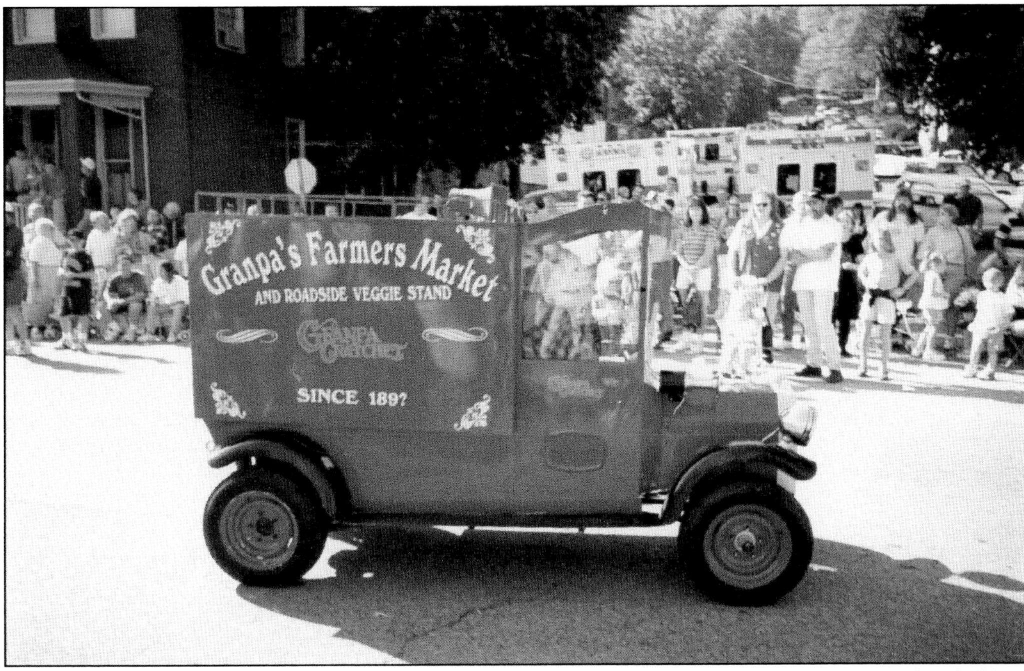

The puppet called Grandpa Crachit is a real favorite with the children. After the parade, children crowd around Grandpa's truck in the children's ride area for a show that allows the children to interact with the puppet. (Largent collection.)

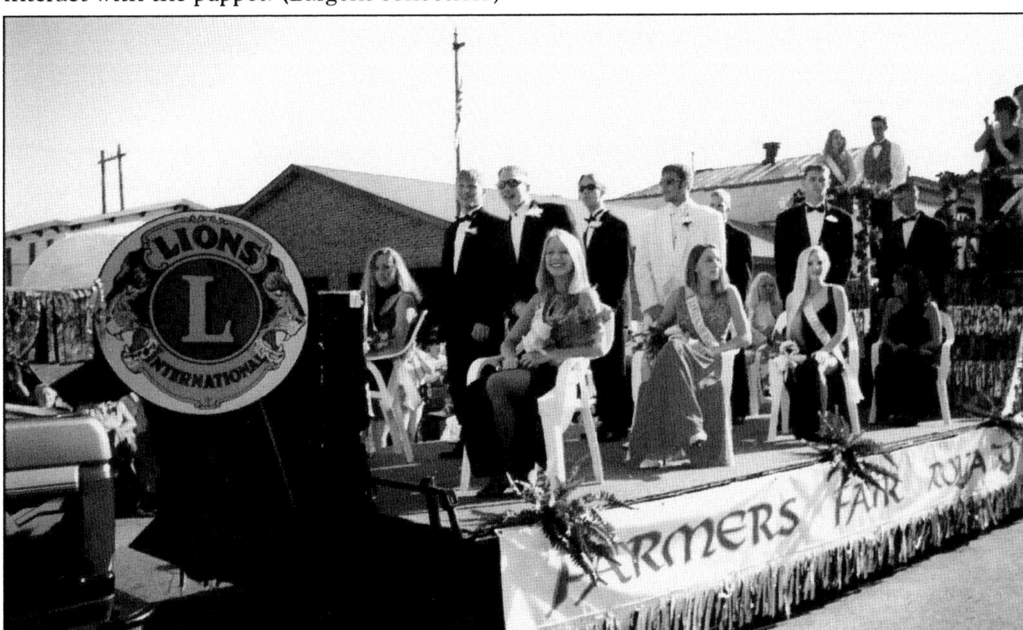

The king and queen contestants of the 1997 farmers fair are shown on the Lions Club float. The queen of the 89th fair was Amy Redmond of South Ripley High School. Batesville High School senior Jon Saner was crowned king. Runners-up crowned were Switzerland County High School senior Cassandra Chase and Dave Gilbert of Southeastern Career Center. (Largent collection.)

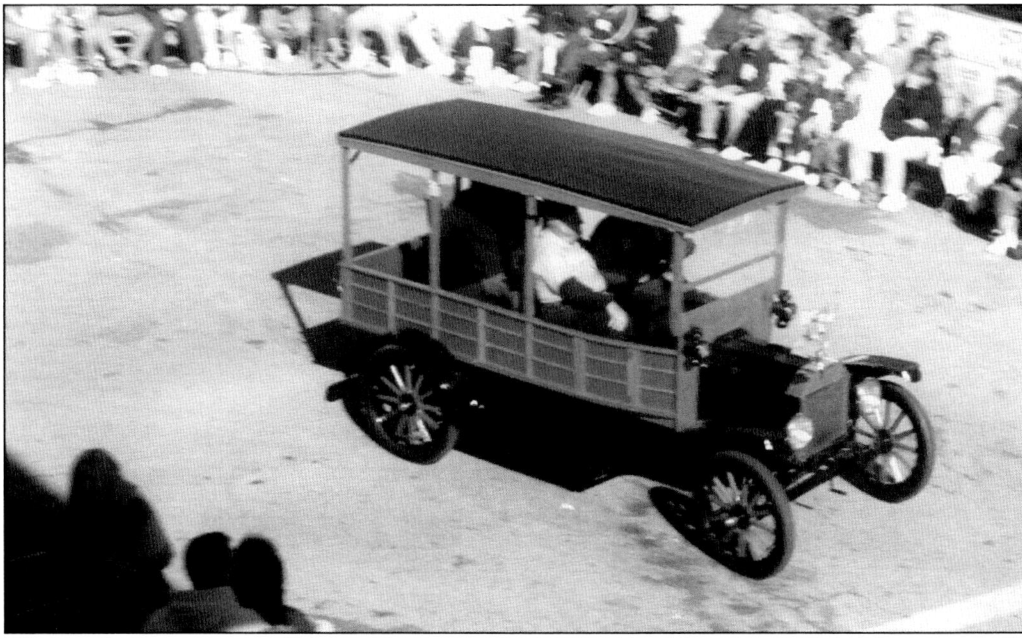

An old wooden Model-T truck owned by Bud Sales turns on Second Street at the 2006 fair. (Largent collection.)

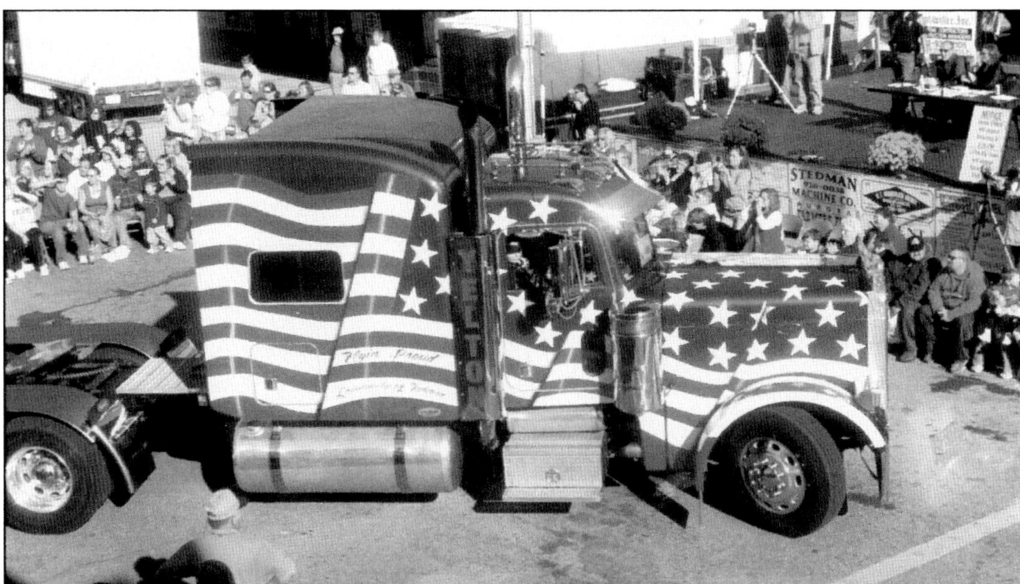

William Yelton of Greendale is the proud owner of this American flag semitruck. Yelton is owner of the Yelton Trucking Company. (Largent collection.)

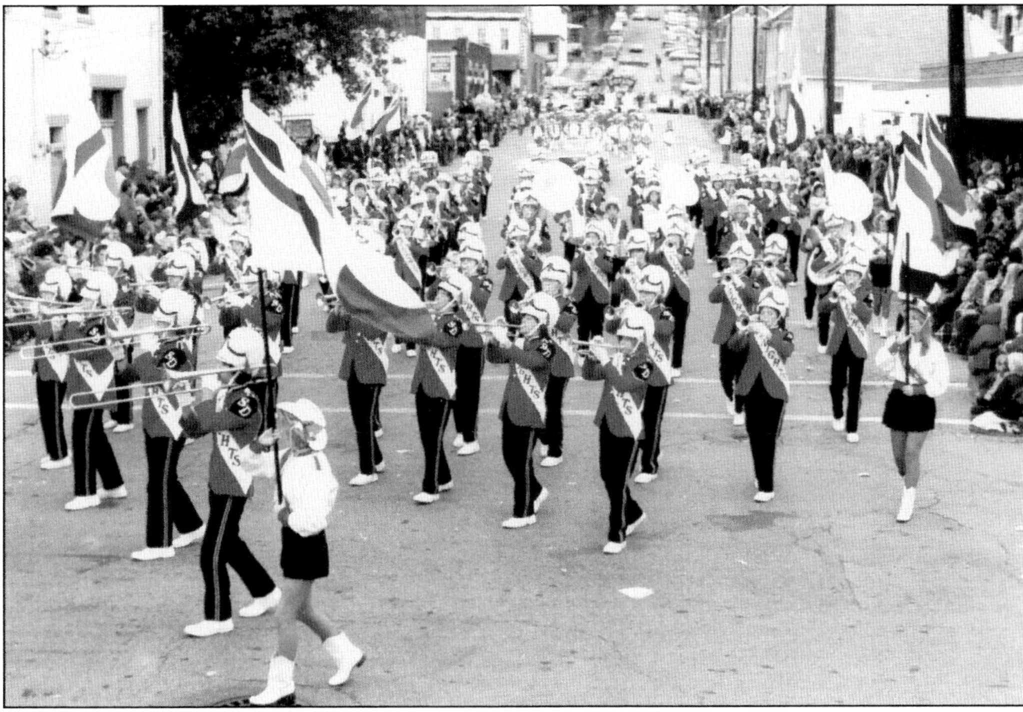

The South Dearborn Marching Band turns the corner in front of the stage. (Tandy collection.)

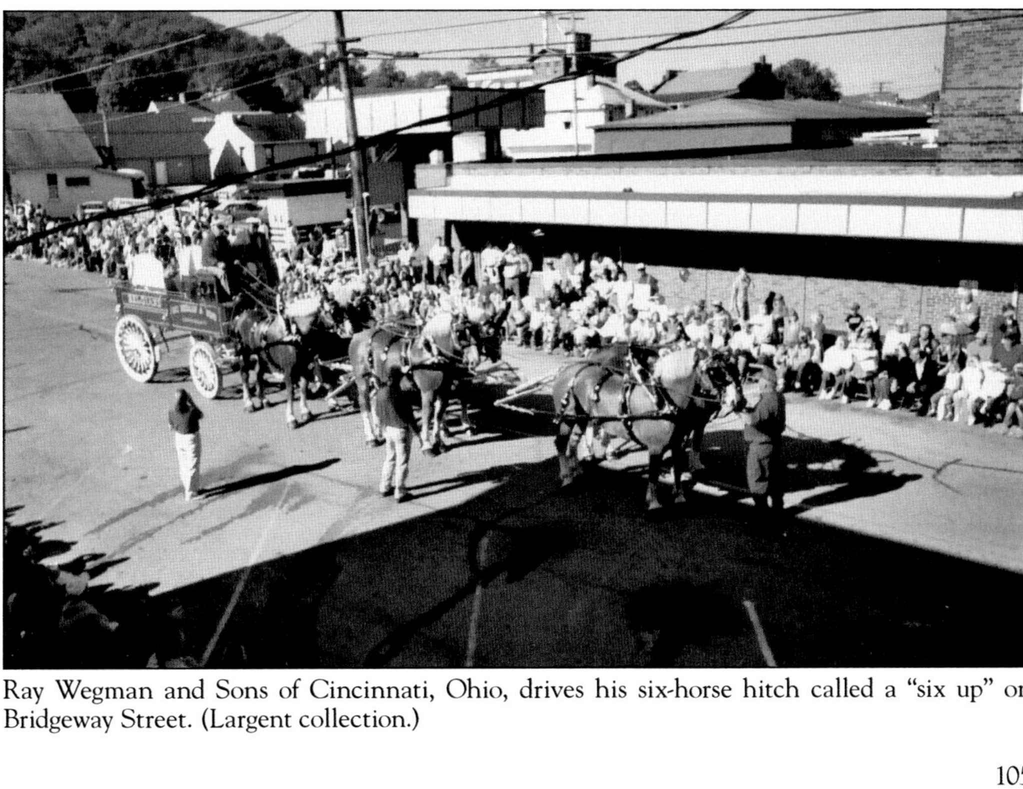

Ray Wegman and Sons of Cincinnati, Ohio, drives his six-horse hitch called a "six up" on Bridgeway Street. (Largent collection.)

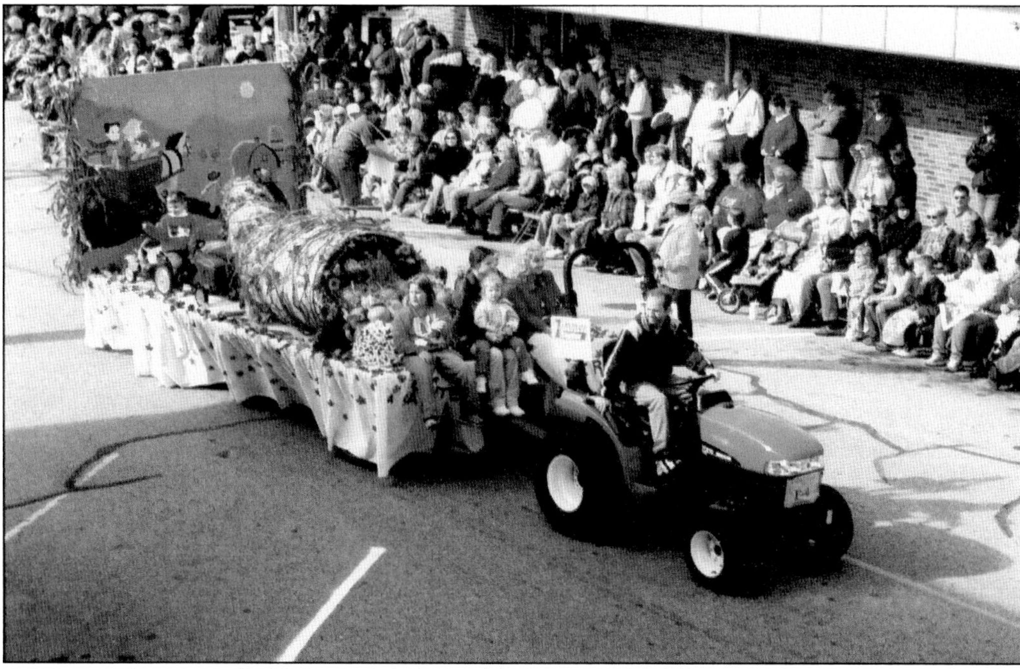

A horn of plenty is pulled by a tractor driven by Steve Kittle. A float similar to this was pulled in the 1948 fair. The tractor was provided by Zimmer Tractor, Aurora. (Largent collection.)

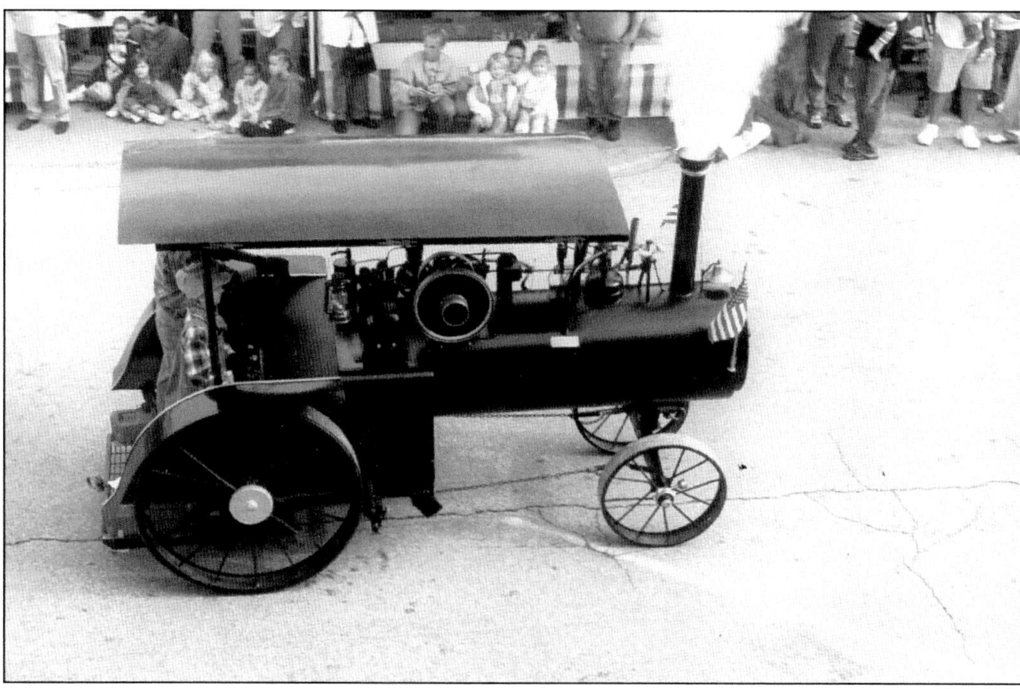

A rare sight for the youngsters, a steam engine built from scrap metal makes its way up Second Street at the 2004 fair. The steam engine is owned by Wendall Minks, Rising Sun, and is being driven by Minks's grandchildren. (Largent collection.)

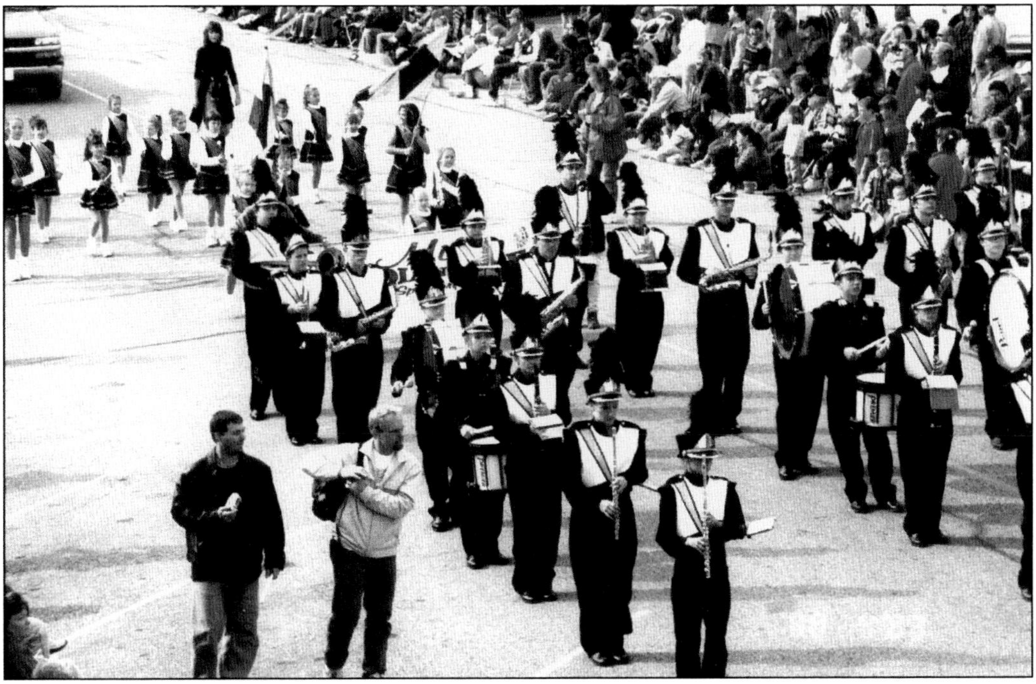

The 2003 Milan High School Band is photographed marching for review in front of the stage. (Largent collection.)

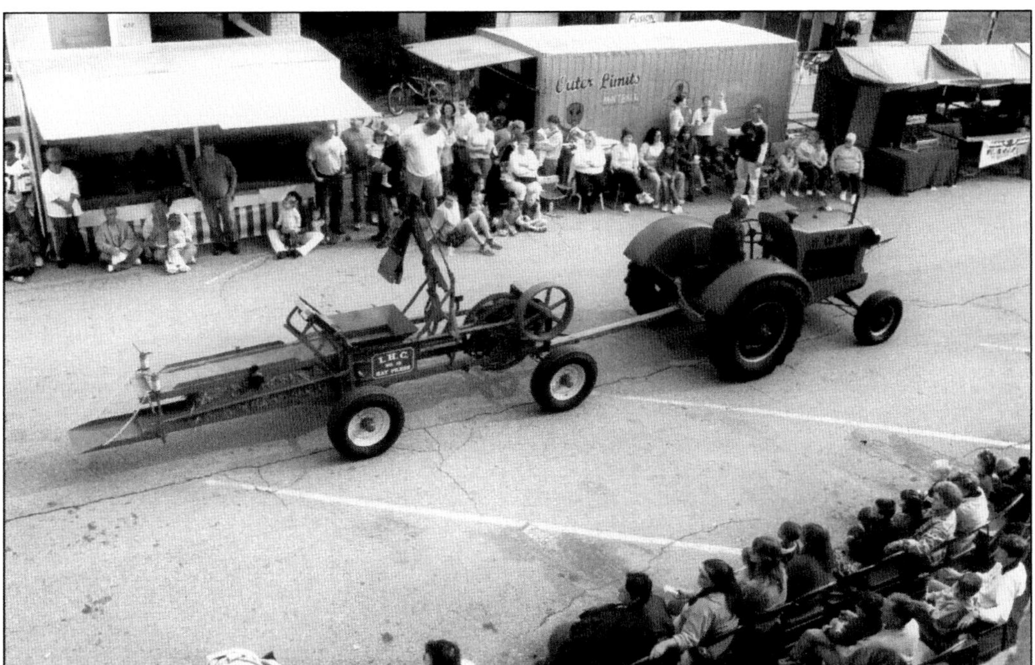

This bird's-eye view is of a restored tractor and hay baler at the 2004 fair. (Largent collection.)

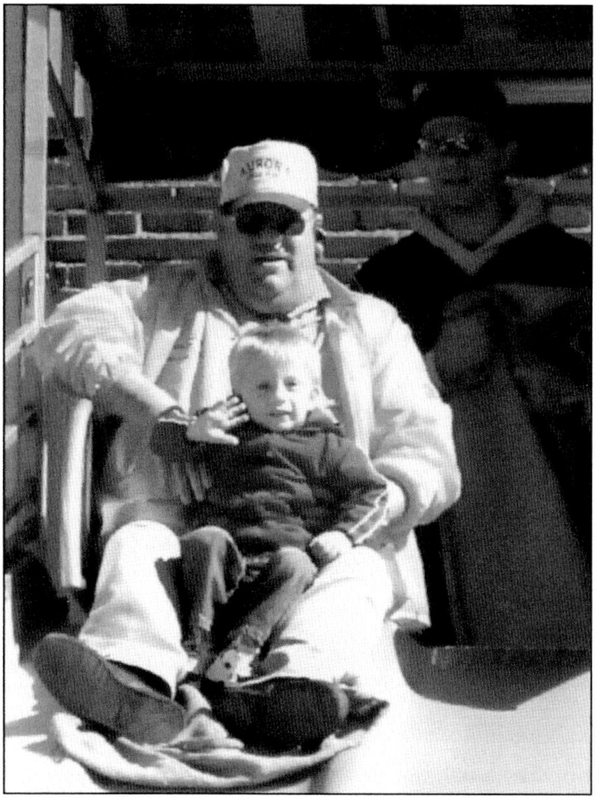

Riding the giant slide is fair committee chairman George Feustel and grandson Logan Richard. Feustel became a member of the Aurora Lions Club in 1987 and joined the fair board in 1988. He is also a past president.

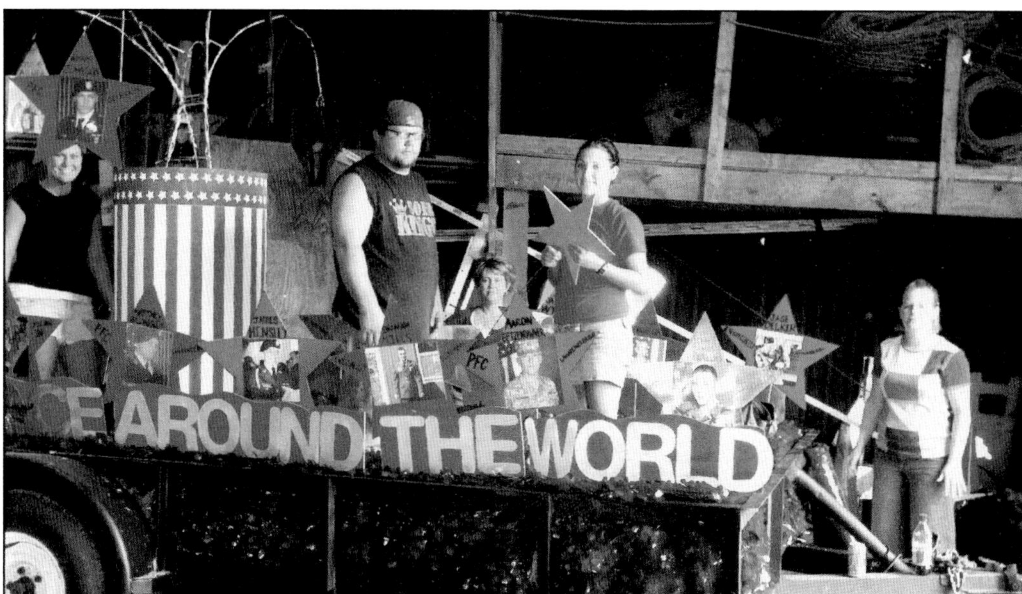

Peace on earth was the theme of the 2007 fair. Employees from United Community Bank build their float in the Aurora City garage. Their float went on to win third for the best business or business organization.

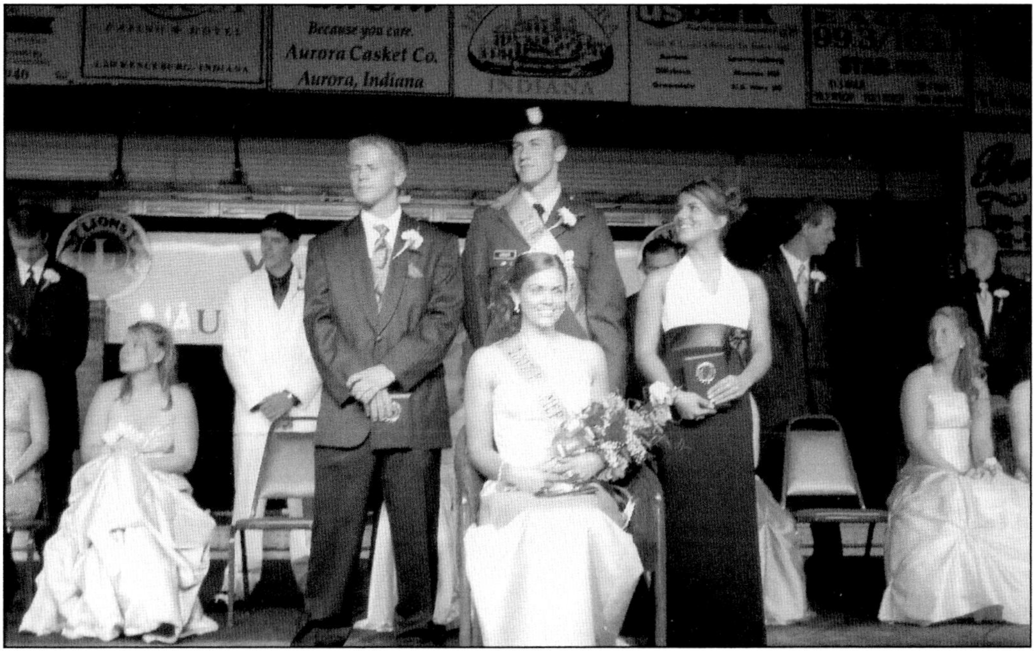

The 2007 Aurora Farmers Fair king is Southeastern Career Center student Nicholas Grider, center, and the queen is Oldenburg Academy student Jillian Knueven, seated. The king runner-up is Mitch Strobl, East Central, and queen runner-up is Brittany Morton. The contest was sponsored by Register Publications and Embarq Communications. (Aurora Lions Club collection.)

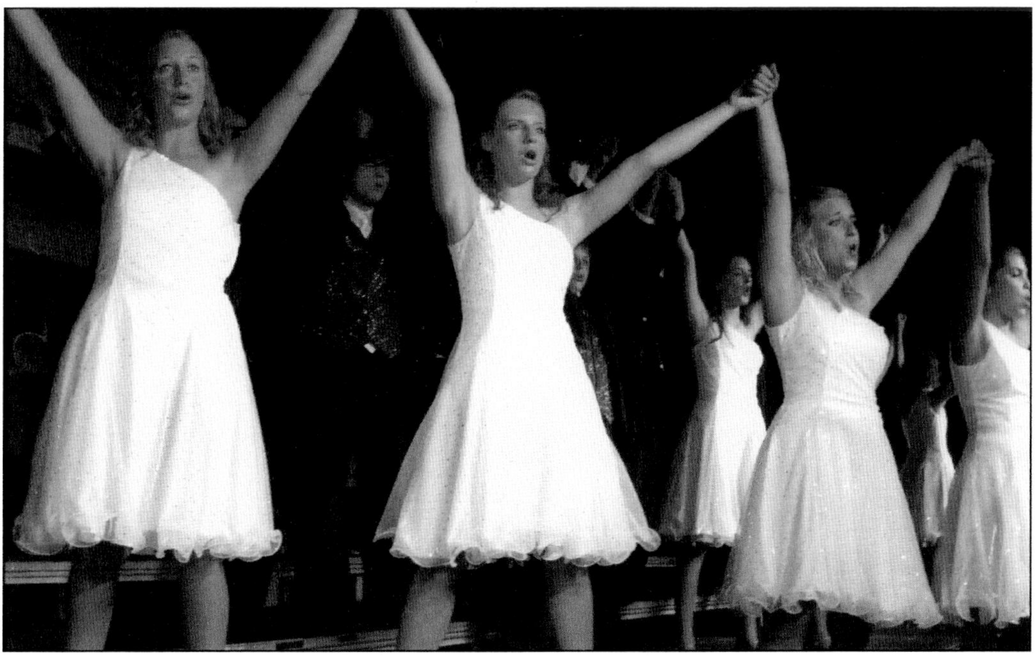

The South Dearborn Opening Knight Show Choir entertained on stage 1 during the queen and king contest. Also entertaining during the event was the South Ripley Young Confederates and the South Dearborn Jazz Band. (Aurora Lions Club collection.)

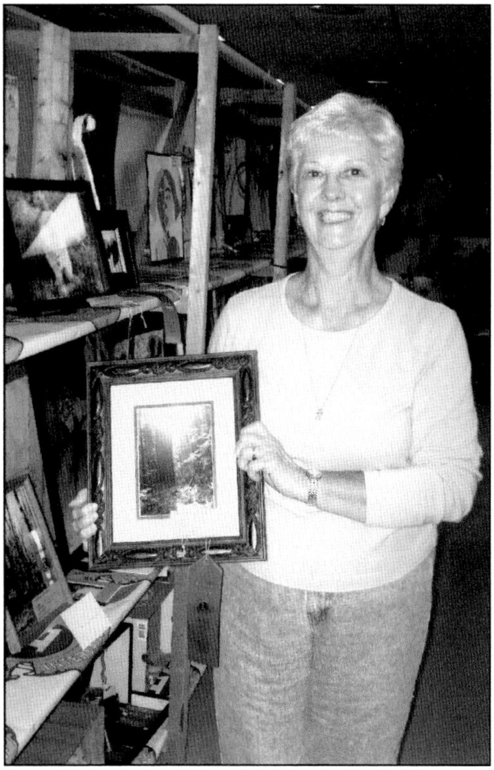

Bonnie Nocks, of Aurora, shows off the red ribbon she won in the photography division. Nocks took the photograph at a recent trip to Northern California. (Aurora Lions Club collection.)

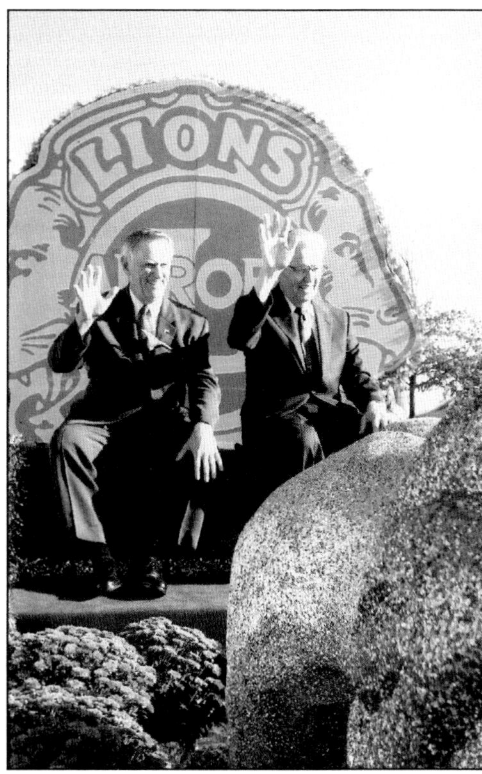

Hometown proud is a great way to describe the grand marshals of the 99th Aurora Farmers Fair, brothers Roger and Millard Rullman. The two Aurora residents owned and managed Rullman Funeral Home in Aurora for many years. In addition to riding on the Aurora Lions Club float in the parade on Saturday, October 6, the two brothers were honored at the main stage. (Aurora Lions Club collection.)

South Dearborn High School principal Rob Moorhead takes a moment behind the stage to organize his notes for the queen and king contest. As a master of ceremony, Moorhead also announced the parade entries from the main stage located on Bridgeway and Second Streets. (Aurora Lions Club collection.)

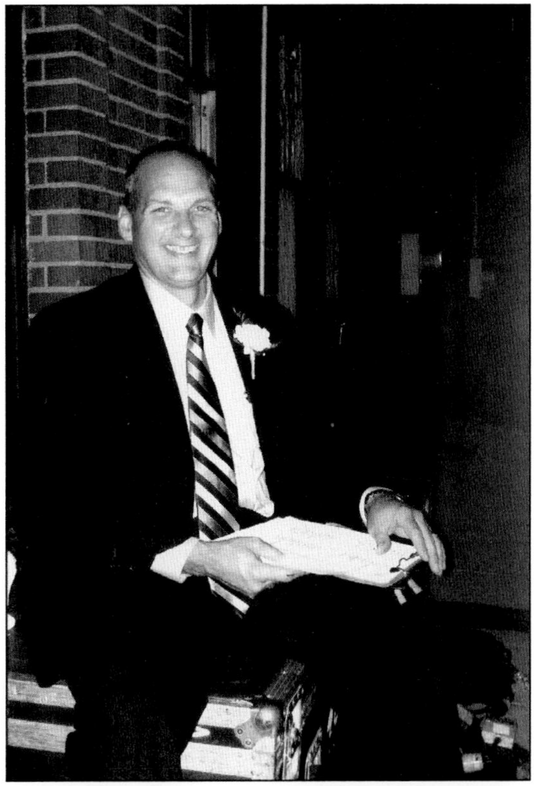

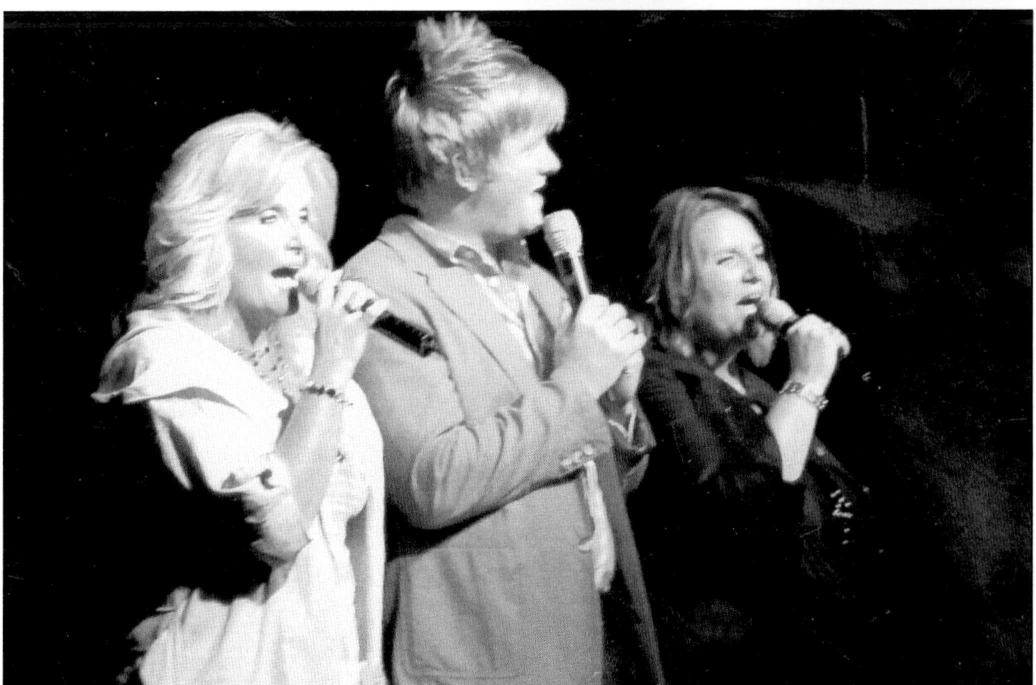

It was standing room only as gospel singers Karen Peck and New River entertained the October 4, 2007, crowd. (Aurora Lions Club collection.)

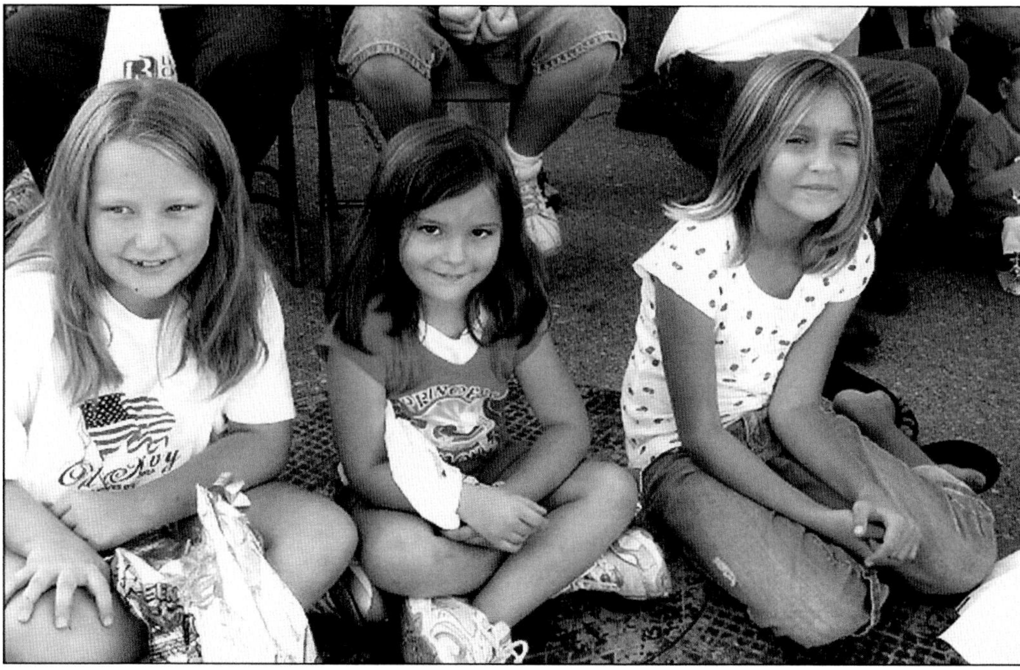

Watching the parade and patiently waiting for candy to be thrown their way are Savanna Ross (left), Rylee Ross (center), and Kennedy Theege. (Aurora Lions Club collection.)

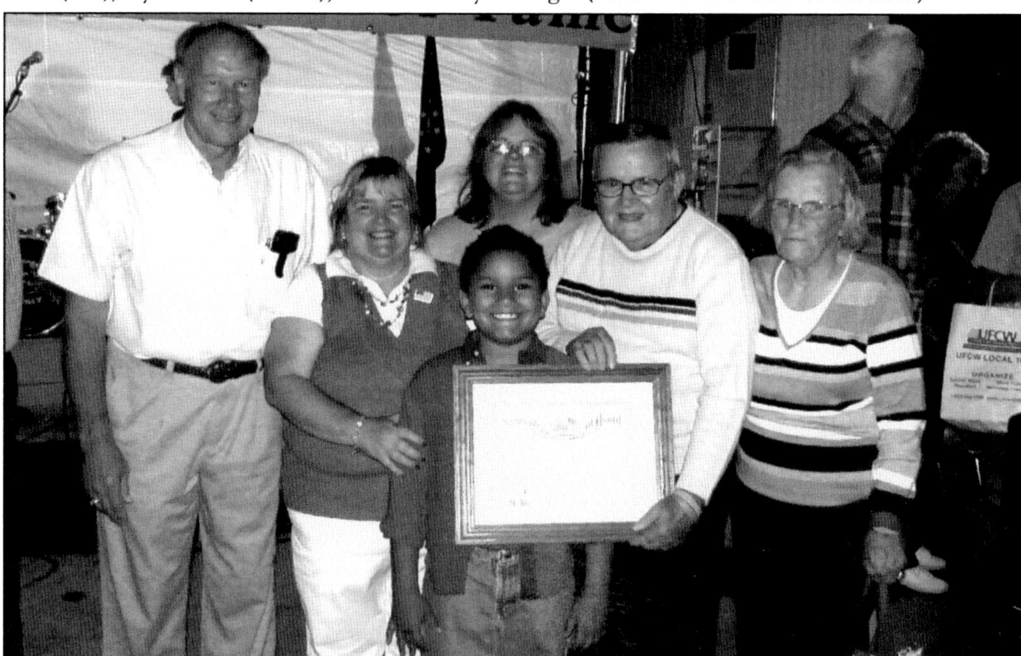

New members were inducted in the Southeastern Indiana Musicians Hall of Fame. From left to right are Indiana 68th District representative Robert Bischoff, daughter Mary Alice Horton, grandson Kobi James, daughter Lora James, inductee Gerald James, and wife Oleva James. Gerald James taught music for 21 years at the Lawrenceburg schools and gives much of his time entertaining senior centers. (Aurora Lions Club collection.)

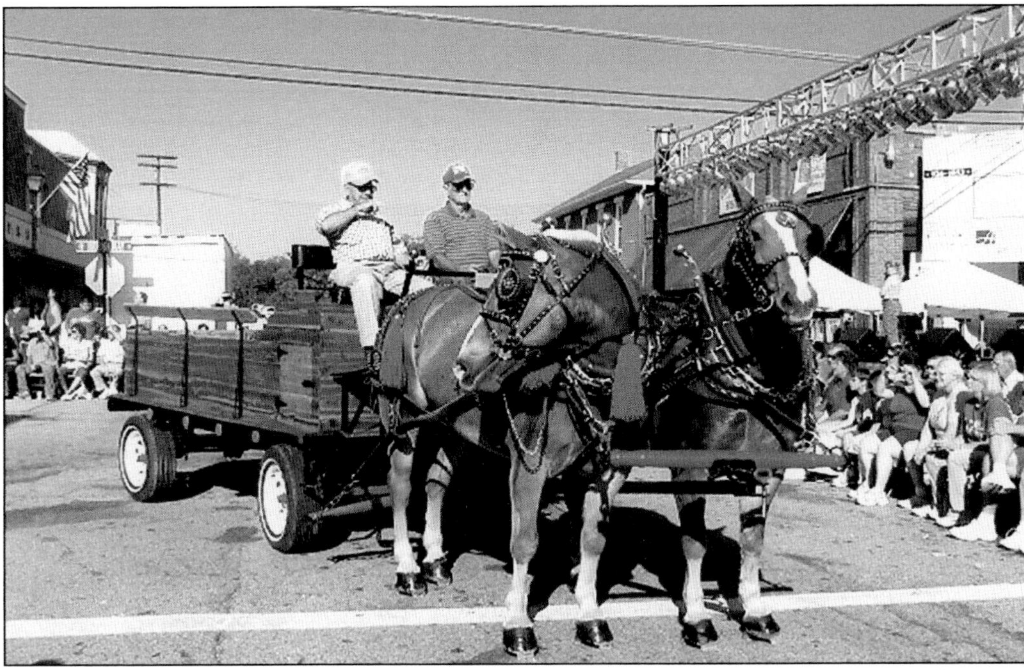

Seated on a pine wagon are Gene Weaver (left), of Aurora, and Daryl Hunt, of Switzerland County. Hunt owns the wagon and team of matching Halflingers. (Aurora Lions Club collection.)

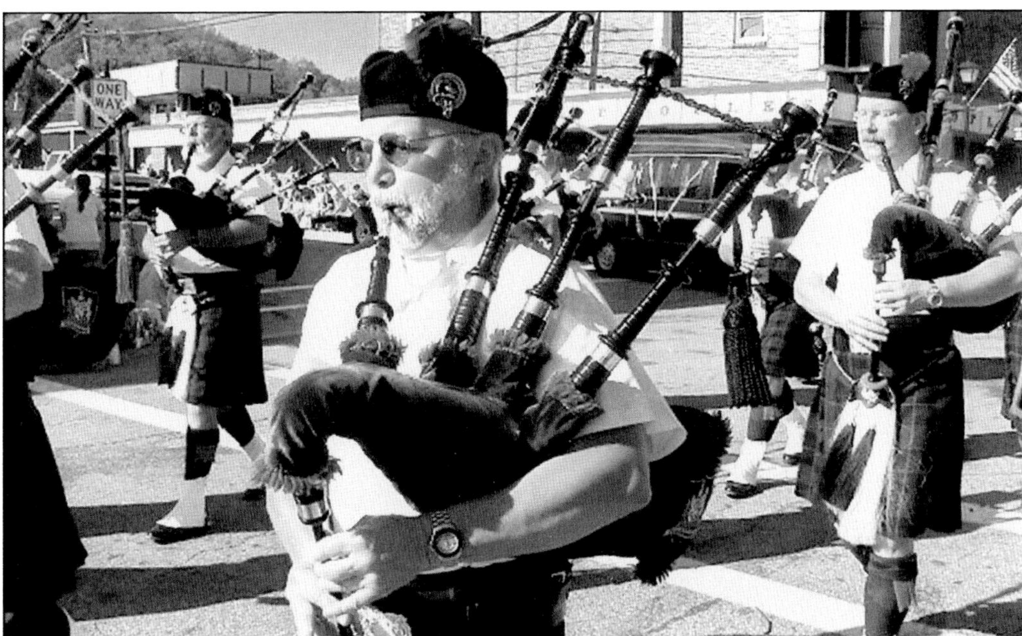

The Cincinnati Caledonians pipe and drum corps entertained the crowd all along the parade route at the 99th annual farmers fair. (Aurora Lions Club collection.)

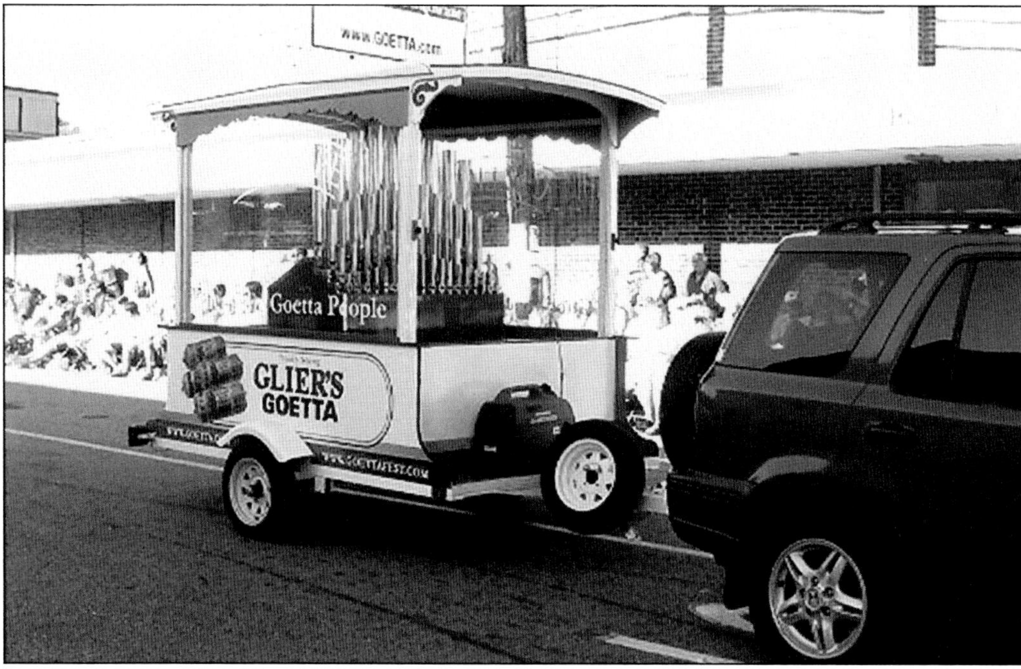

Loud pipe music played by the Glier's Goetta calliope makes a stop on Bridgeway Street in front of the stage. Since the late 1960s, Glier's Goetta has made its home in a historic building on Pike Street in Covington, Kentucky. (Aurora Lions Club collection.)

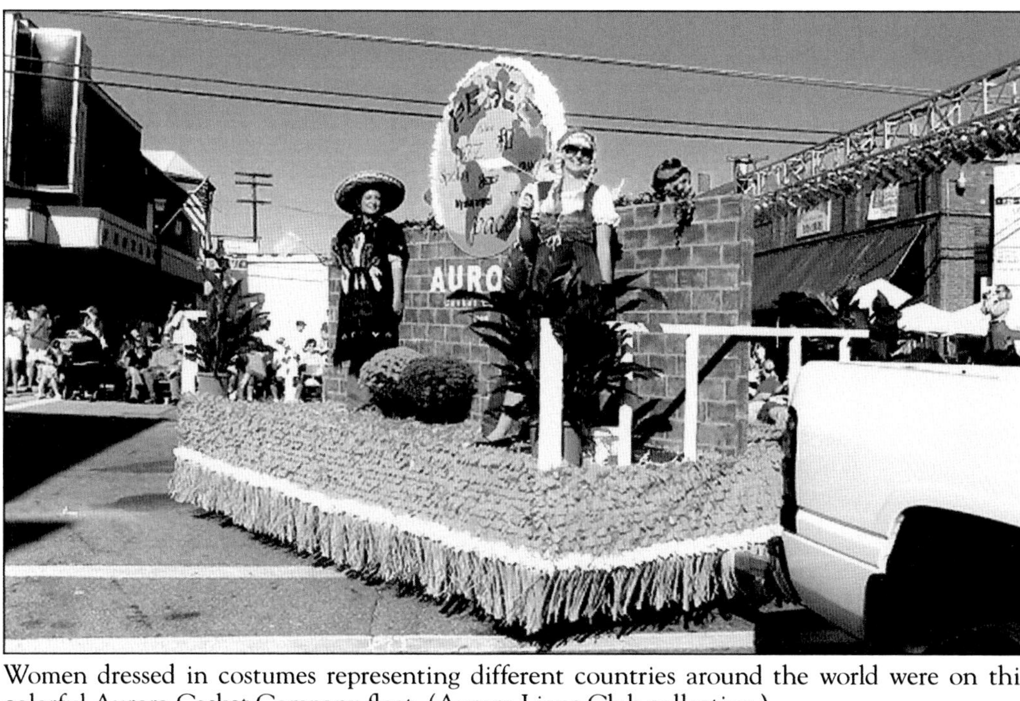

Women dressed in costumes representing different countries around the world were on this colorful Aurora Casket Company float. (Aurora Lions Club collection.)

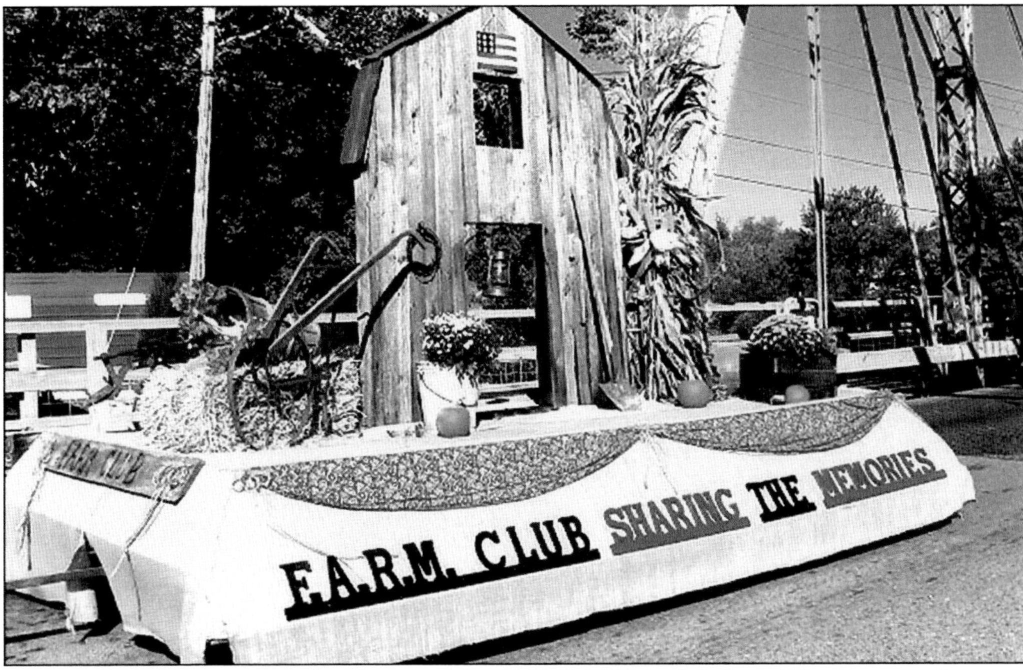

Sharing the Memories was the theme of the 2007 F.A.R.M. club float. The Farming Antiques and Related Machinery (F.A.R.M.) has over 200 members. In the background is the George Street Bridge. The parade committee stages the various parade exhibits on the east side of the bridge. (Aurora Lions Club collection.)

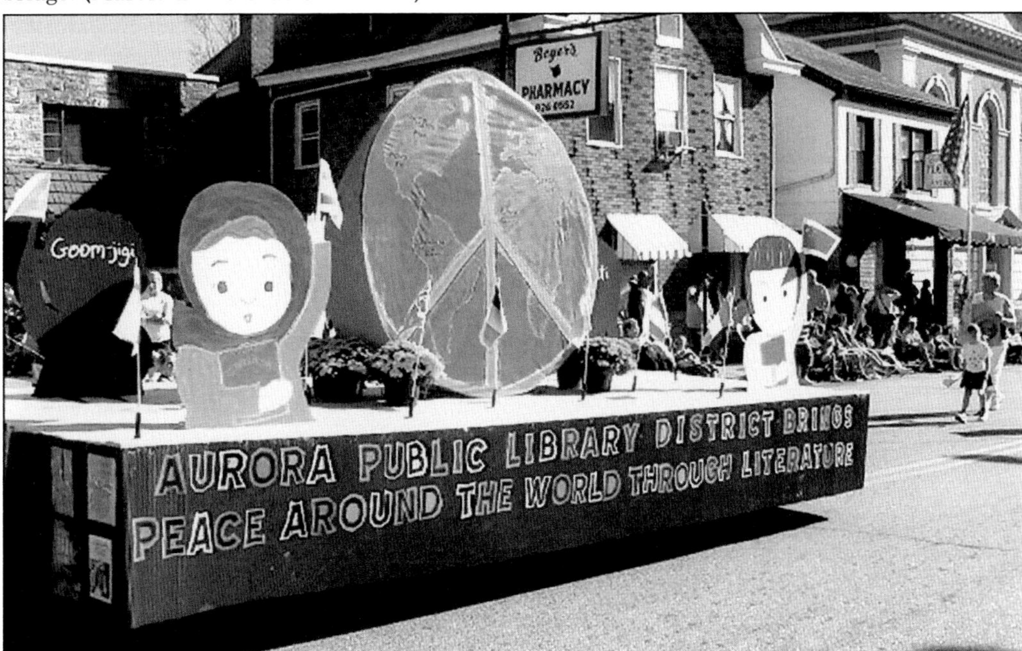

Peace Around the World Through Literature was the theme of the 2007 Aurora Public Library District. This photograph was taken on Third Street in front of Beyers Pharmacy. (Aurora Lions Club collection.)

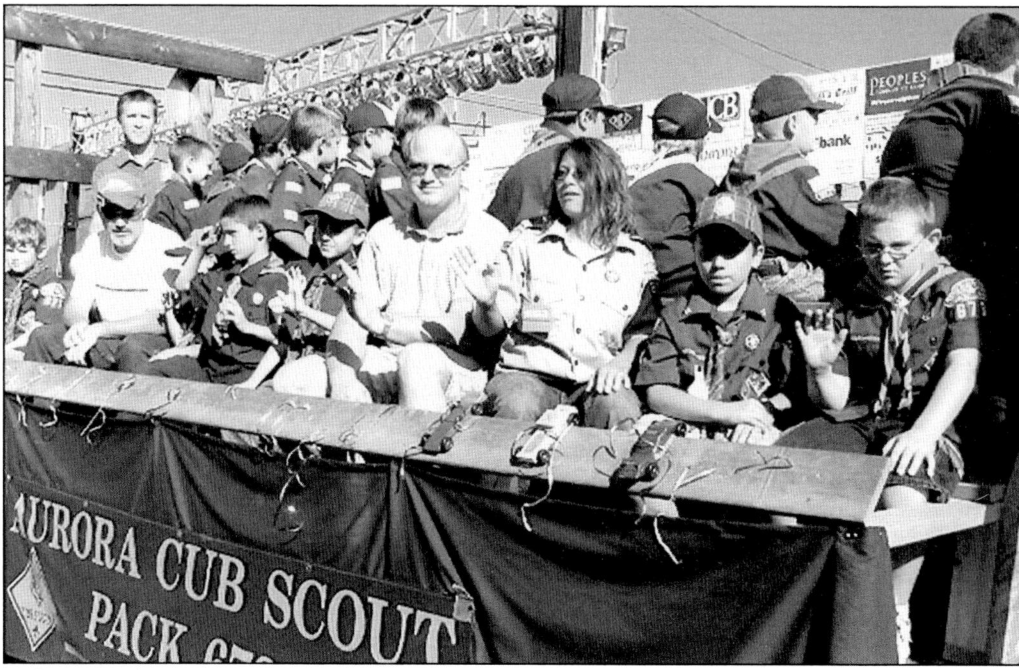

Cub Scouts from Aurora Cub Scout Pack 673 wave to the crowd along the parade route. (Aurora Lions Club collection.)

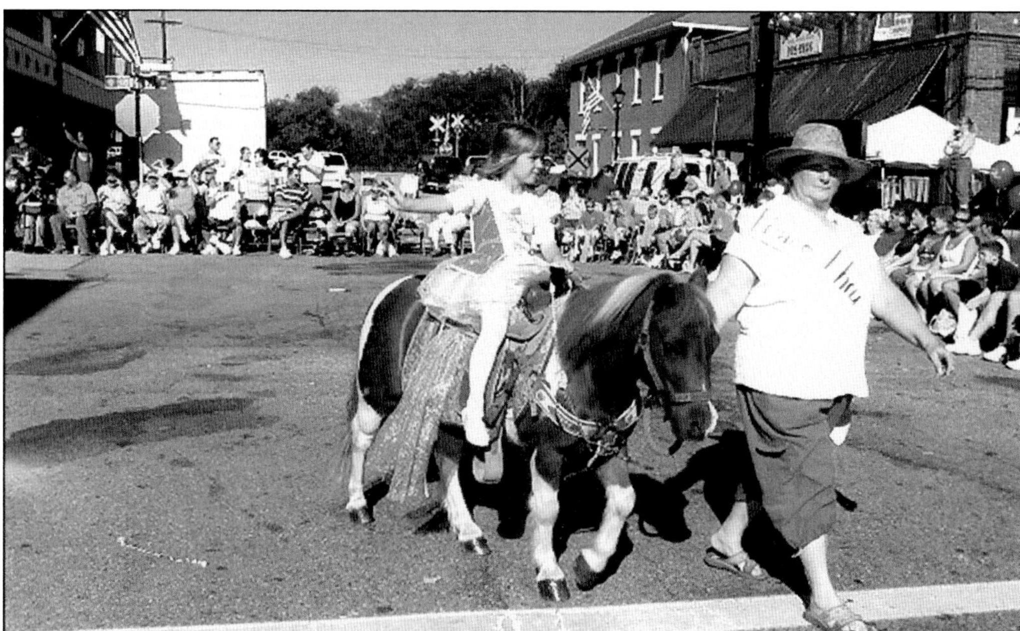

A princess on her steed promotes the parade theme Peace on Earth. (Aurora Lions Club collection.)

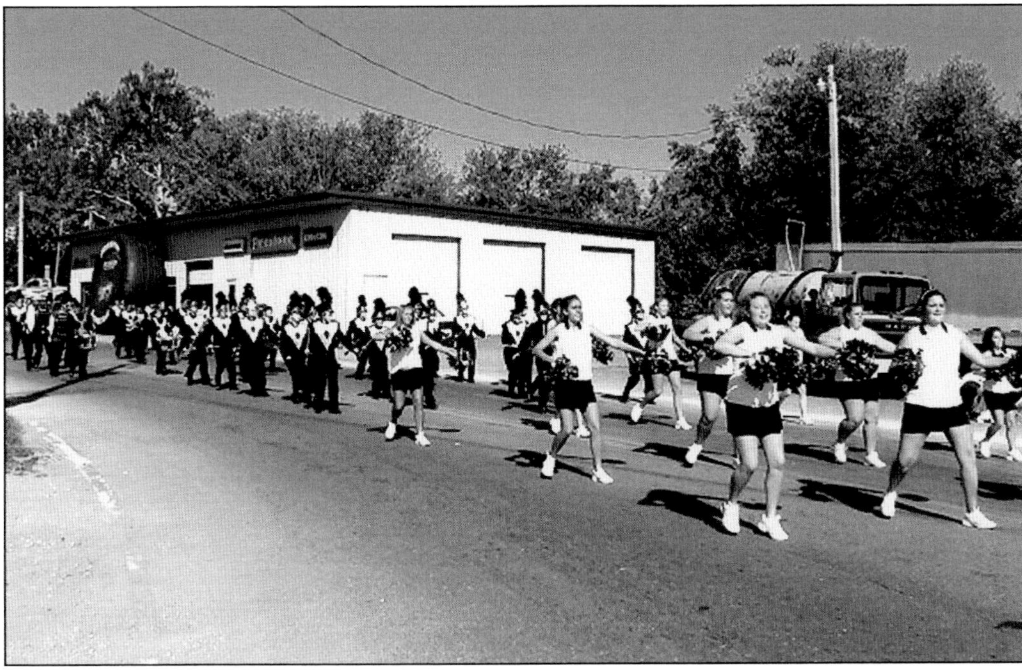

The Milan High School pom-pom girls lead the band at the beginning of the parade route in front of Aurora Tire. Milan High School is located in Milan, Indiana. (Aurora Lions Club collection.)

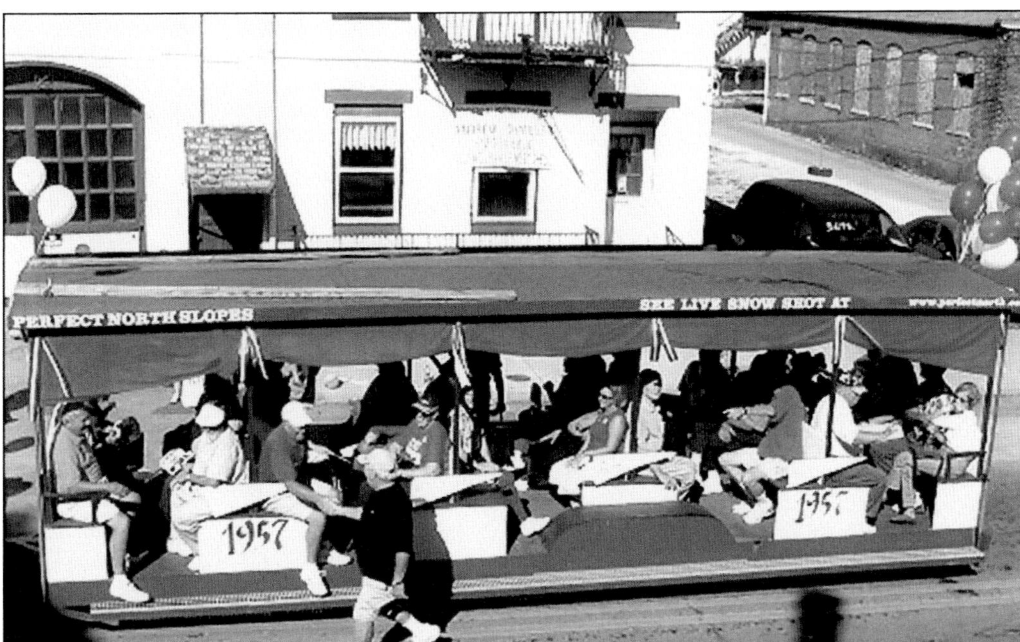

The Aurora High School class of 1957 celebrates its 50th year with this streetcar-style float. The class received a second place in the best youth group, school class, or organization. This photograph was taken from the balcony of the Applewood Restaurant. (Aurora Lions Club collection.)

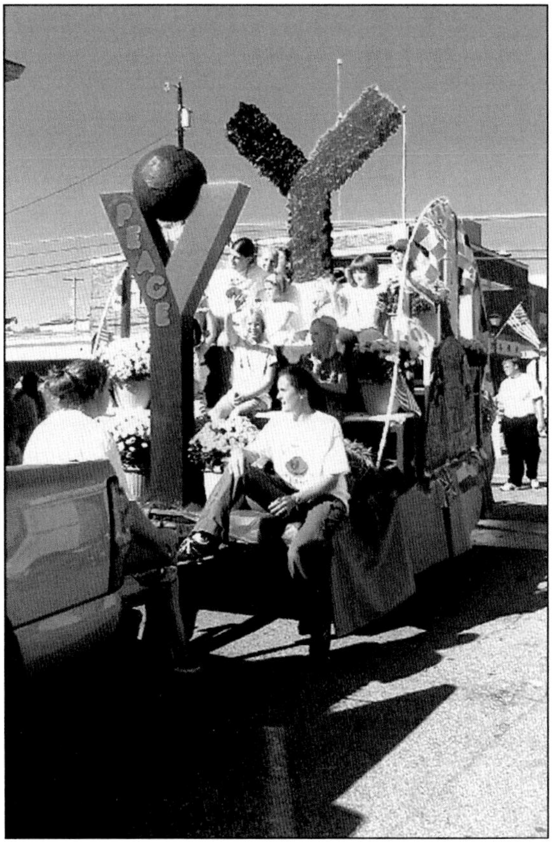

The Dearborn County YMCA won a second with its colorful float. (Aurora Lions Club collection.)

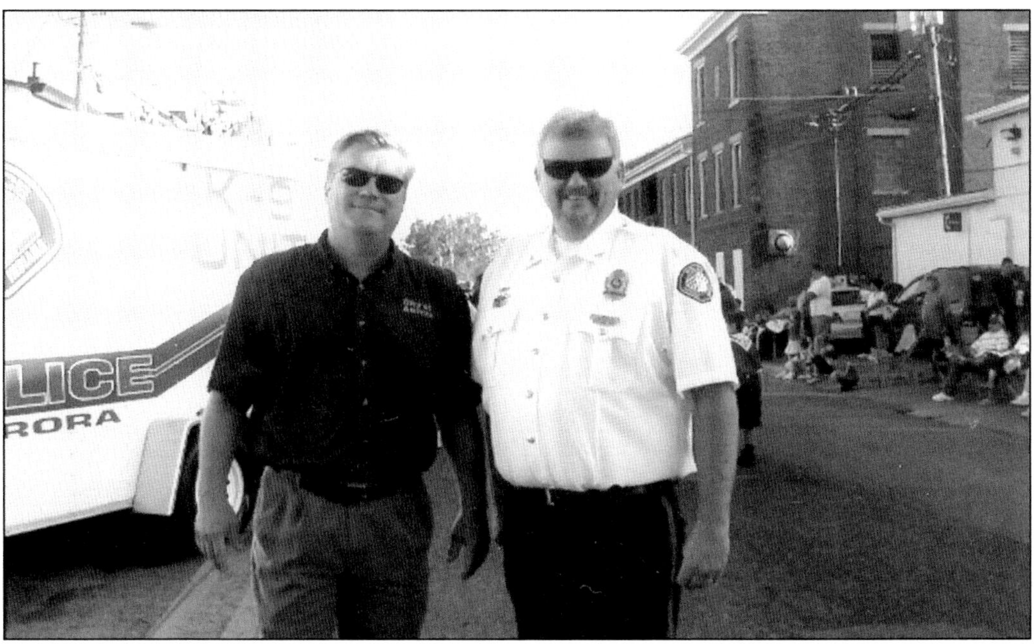

Mayor Donnie Hastings and police chief Dana Cotton led the 99th farmers fair parade. (Aurora Lions Club collection.)

Bob the Realtor is a float based on the television cartoon *Bob the Builder*. From left to right are (first row) Brooklyn Allen, Vern Walts, Olivia Lamb, and Jan Fehrman; (second row) owner of Fehrman Realtors Bob Fehrman and Josh Mangold dressed as the character Spudz. The float's theme was "Can we sell it, Yes we can." (Aurora Lions Club collection.)

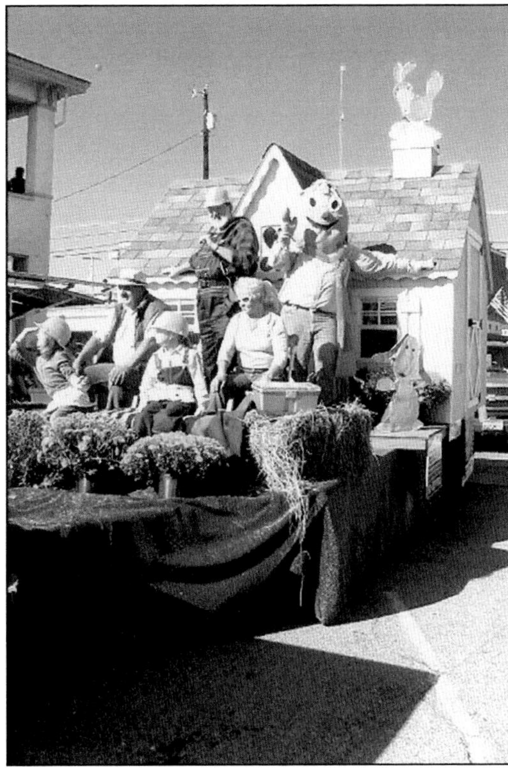

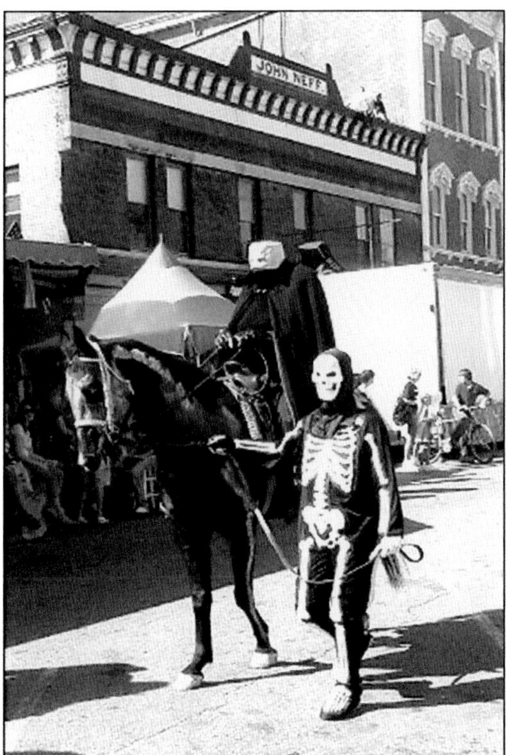

A headless horseman led by a skeleton was one of several scary characters promoting the Ohio County Historical Society event Nightmare on Main Street. The group received a second place for best adult organization. The child-friendly event is held every October in Rising Sun. (Aurora Lions Club collection.)

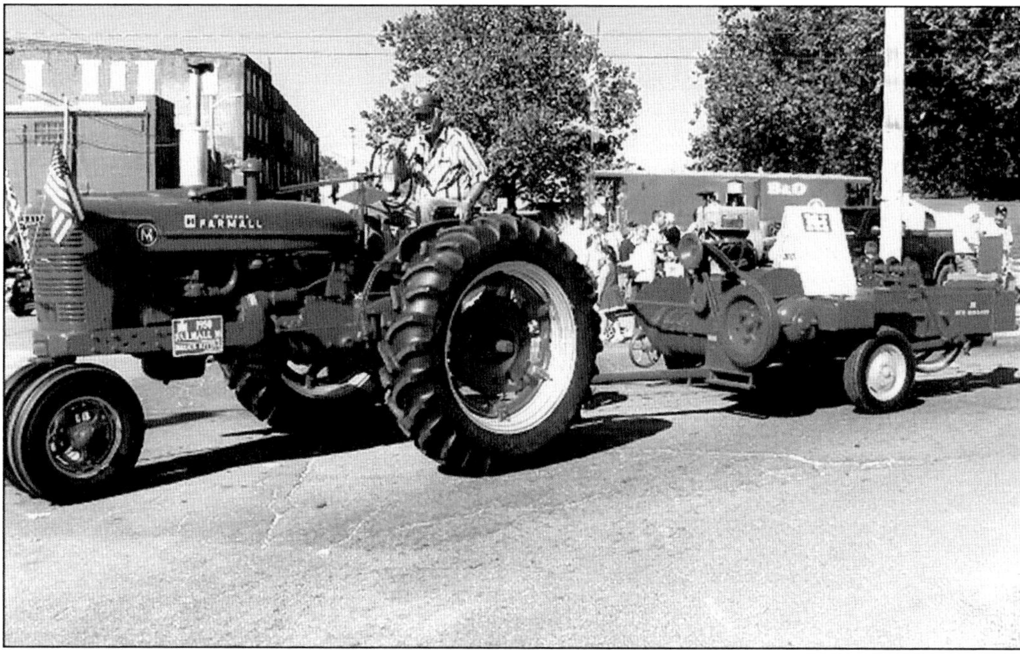

Allen Bunger, of Rising Sun, is driving this antique 1950 Farmall M tractor by International Harvester. Also in the parade (not shown) was his cousin Bruce Kittle driving a 1928 Farmall Regular. Both tractors were purchased new and originally owned by their grandfather Carl Kittle. Bunger is pulling a 1953 New Holland baler. (Aurora Lions Club collection.)

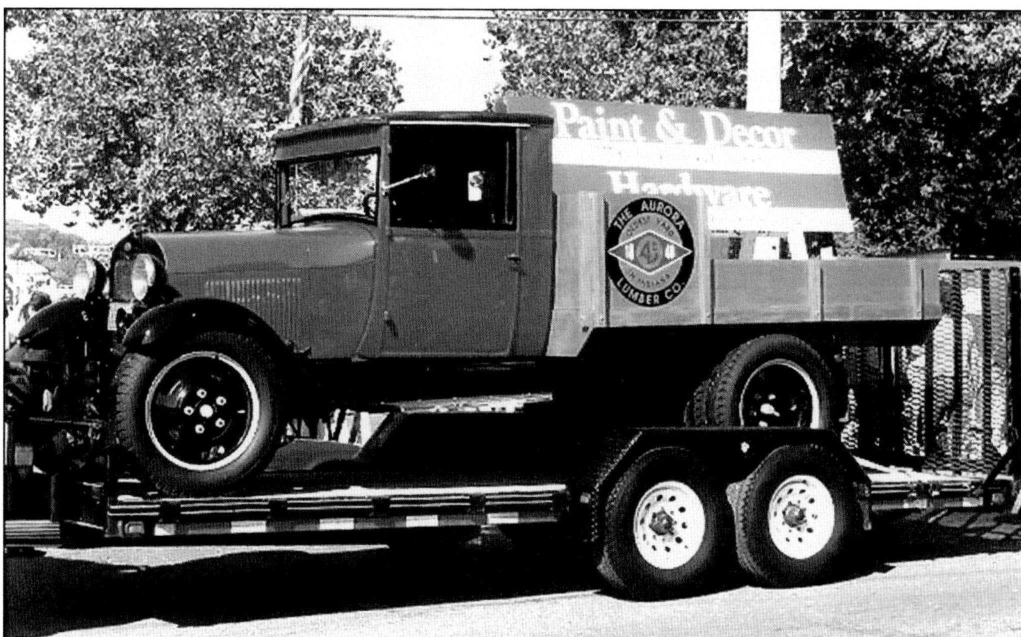

This 1930 Ford 1.5-ton display is owned by the Aurora Lumber Company. It was originally owned by Fred Evans of Aurora. Evans sold it to Theo Beckman, and it sat in front of his Sunoco Station for several years. Carl and Larry Petty then purchased it and had it restored to its present state. (Aurora Lions Club collection.)

The Teddy Bear was a favorite ride in the children's area. (Aurora Lions Club collection.)

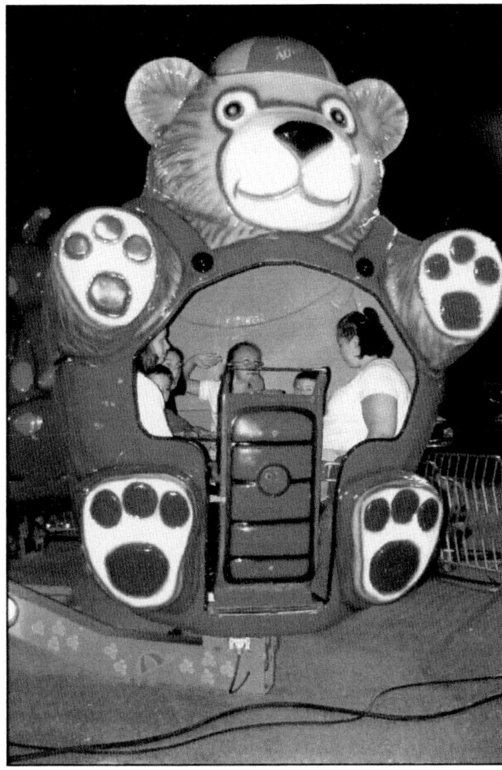

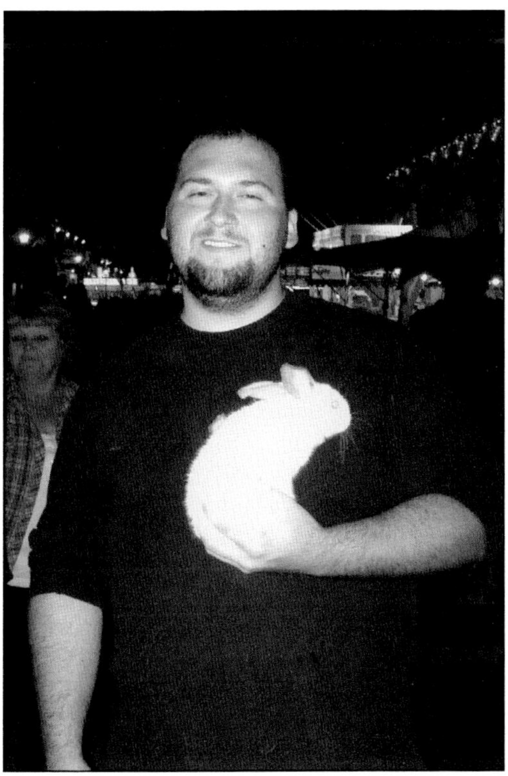

C. J. Allen, of Rising Sun, won a rabbit at one of the game booths. (Aurora Lions Club collection.)

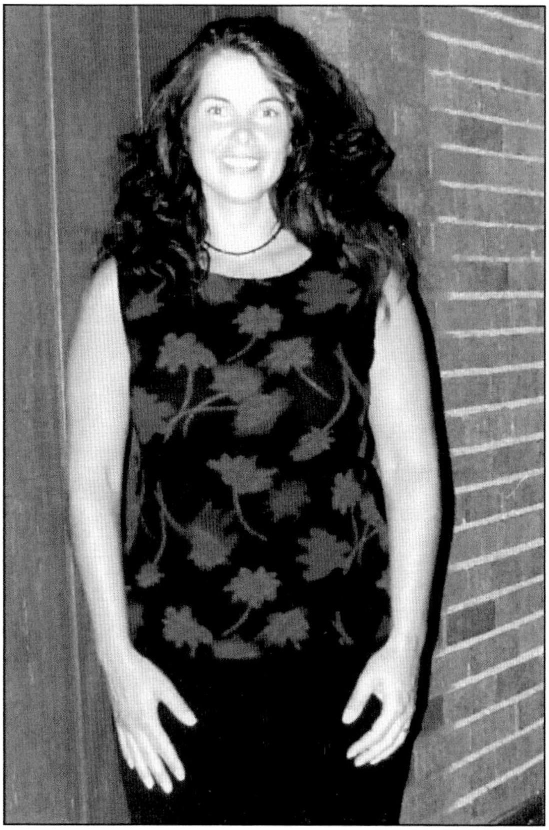

Singing entertainer and hometown girl Allison Kyle sang on the main stage on Thursday, October 4, 2007. Kyle was the 1988 farmers fair queen. (Aurora Lions Club collection.)

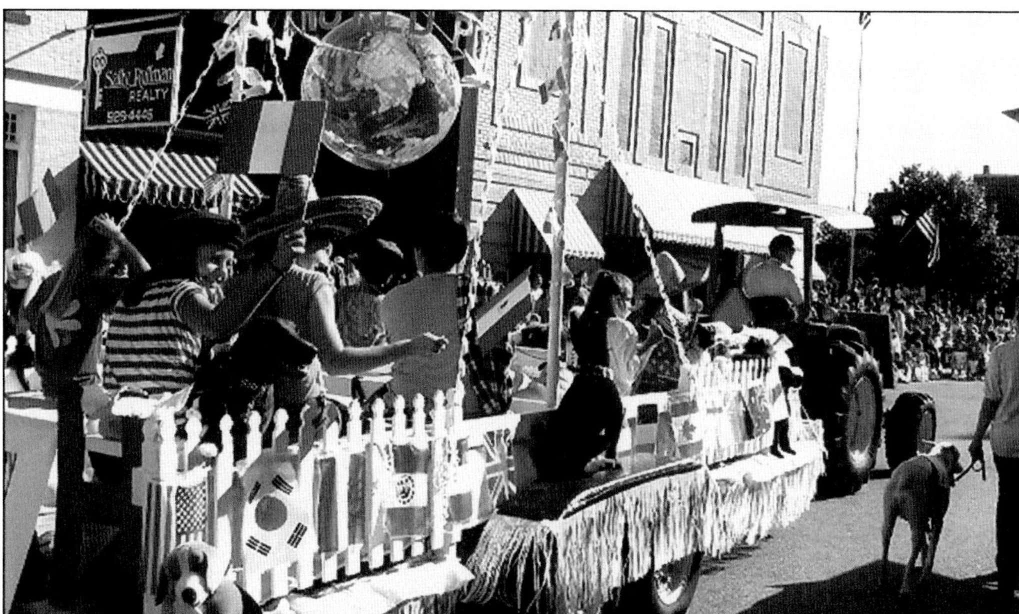

Winning first for the best adult organization was the Partners for Animal Welfare Society (PAWS) animal rescue organization. (Aurora Lions Club collection.)

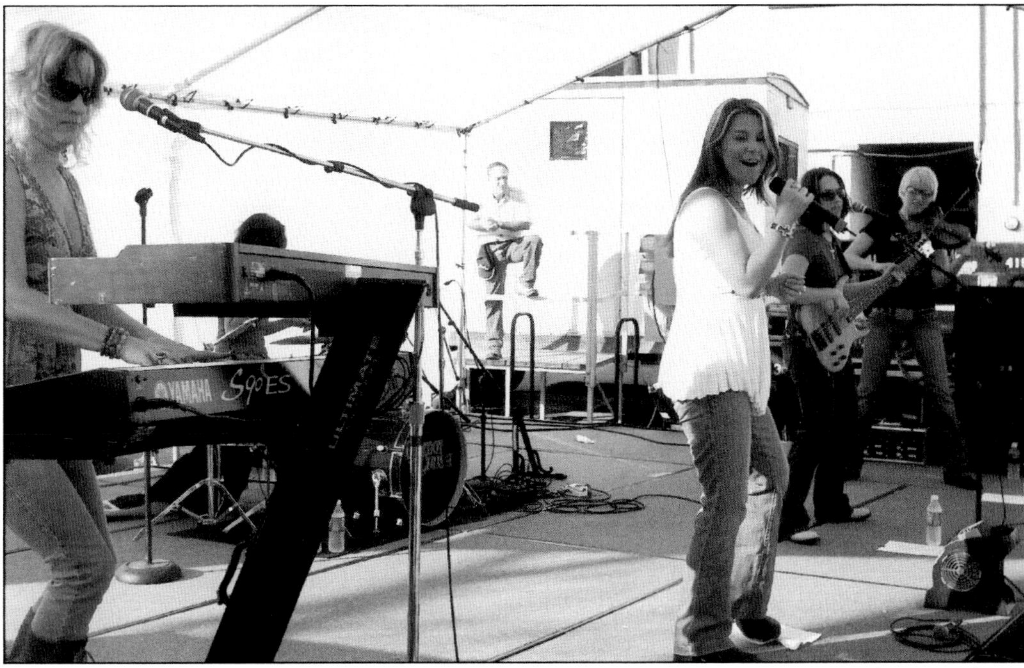

The all-female group Cowboy Crush entertained the crowd on stage 2 on Saturday, October 6, 2007. (Aurora Lions Club collection.)

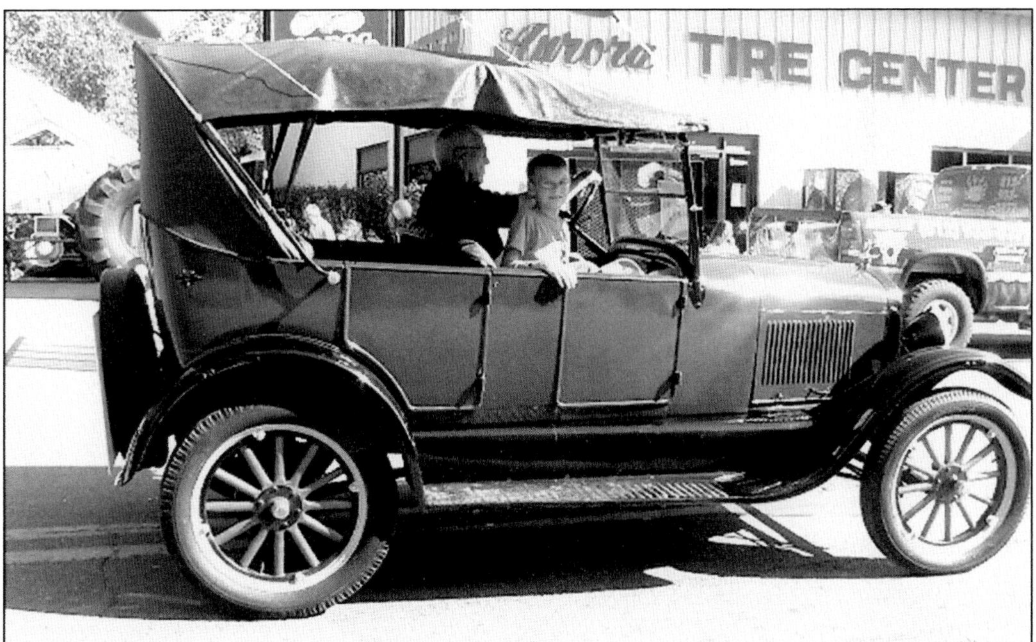

A restored antique car starts the parade route in front of Aurora Tire Center on the corner of Main Street and State Route 56. (Aurora Lions Club collection.)

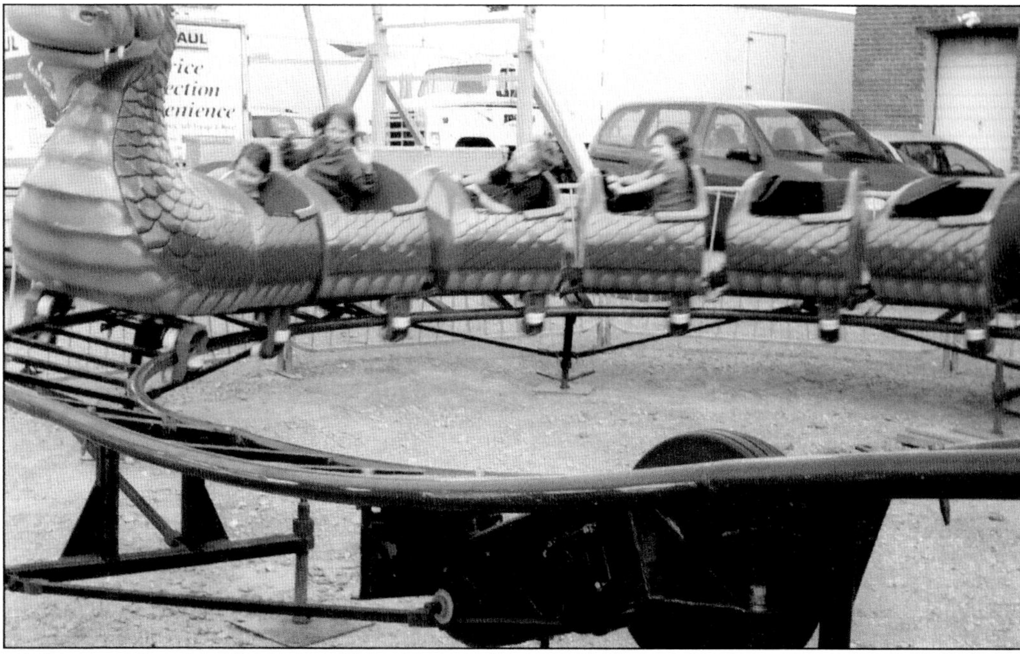

Children ride the dragon roller coaster located in the children's area. (Aurora Lions Club collection.)

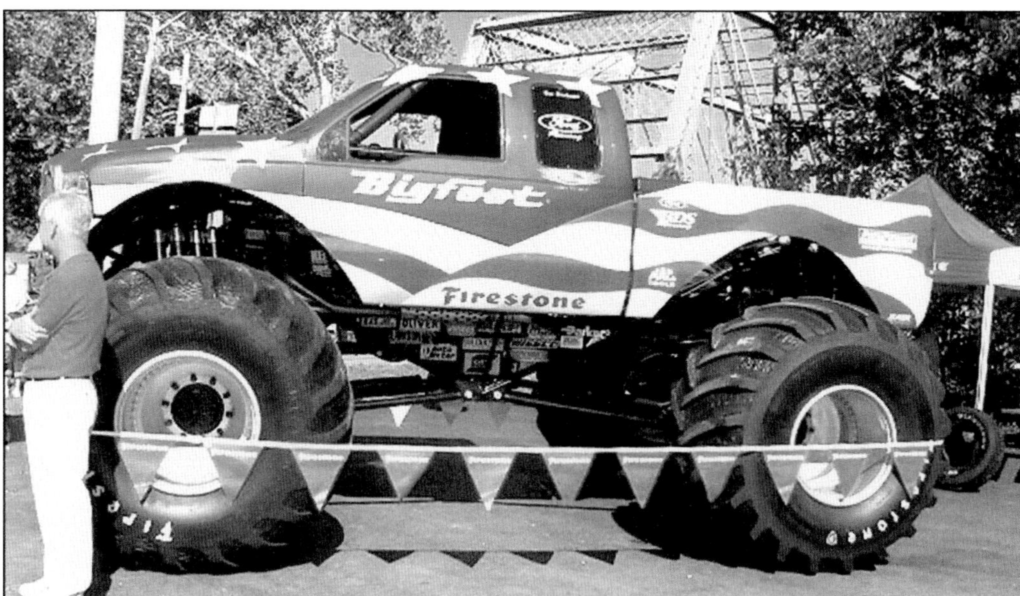

The truck called Bigfoot, sponsored by the Firestone Company, was on exhibit at the Aurora Tire Center. (Aurora Lions Club collection.)

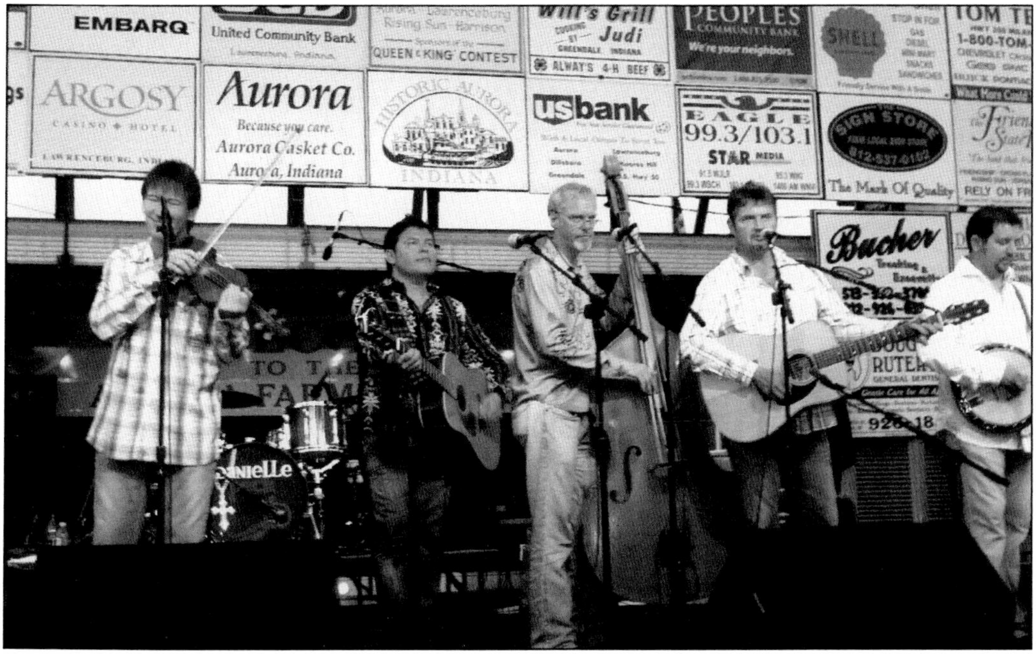

Bluegrass group the Grascals entertained on the main stage on Saturday. Aurora was their first stop after winning the Bluegrass 2007 Entertainer of the Year award for the second consecutive year. (Aurora Lions Club collection.)

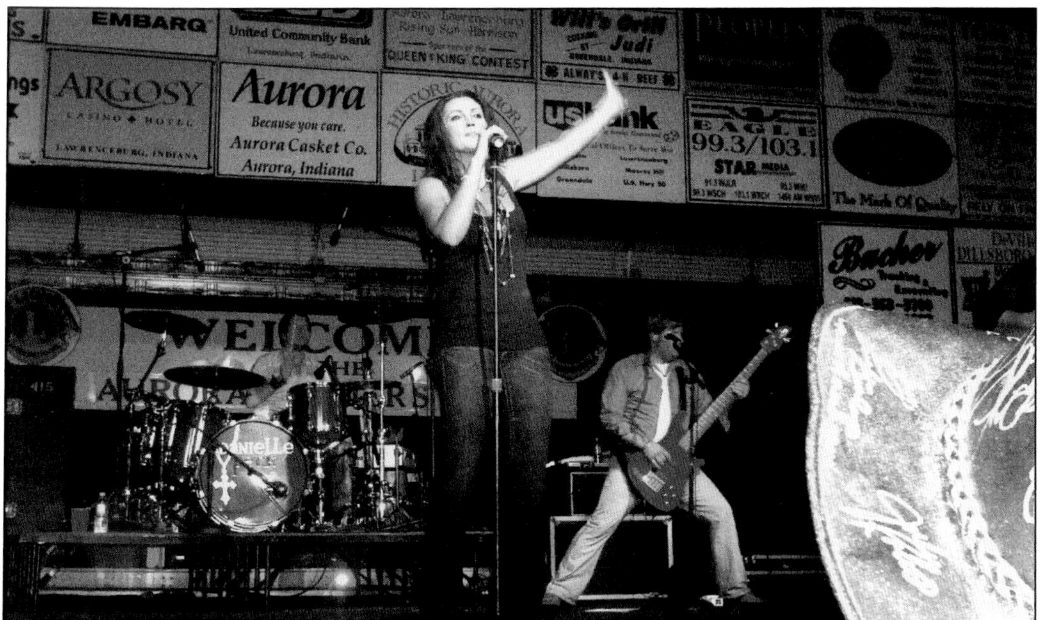

Nominated for country-and-western Female Artist of the Year, Danielle Peck entertained the Saturday night crowd on stage 1. Peck grew up in Coshocton, Ohio. Her single release "I Don't" is a hit and still climbing. (Foutch collection.)

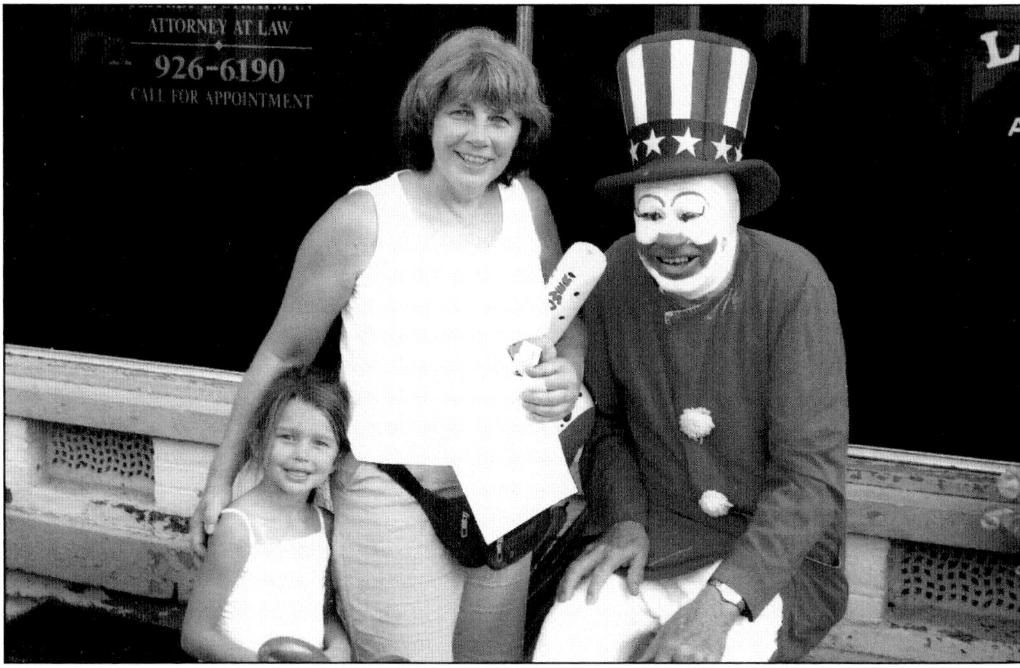

Quacky the Clown takes a moment out of his busy balloon-making schedule to pose with Jenny Awad, coauthor of this book, and her four-year-old granddaughter McKenna Murray-Fehr. Quacky has entertained at the fair for over 30 years and was grand marshal in 2006. (Aurora Lions Club collection.)

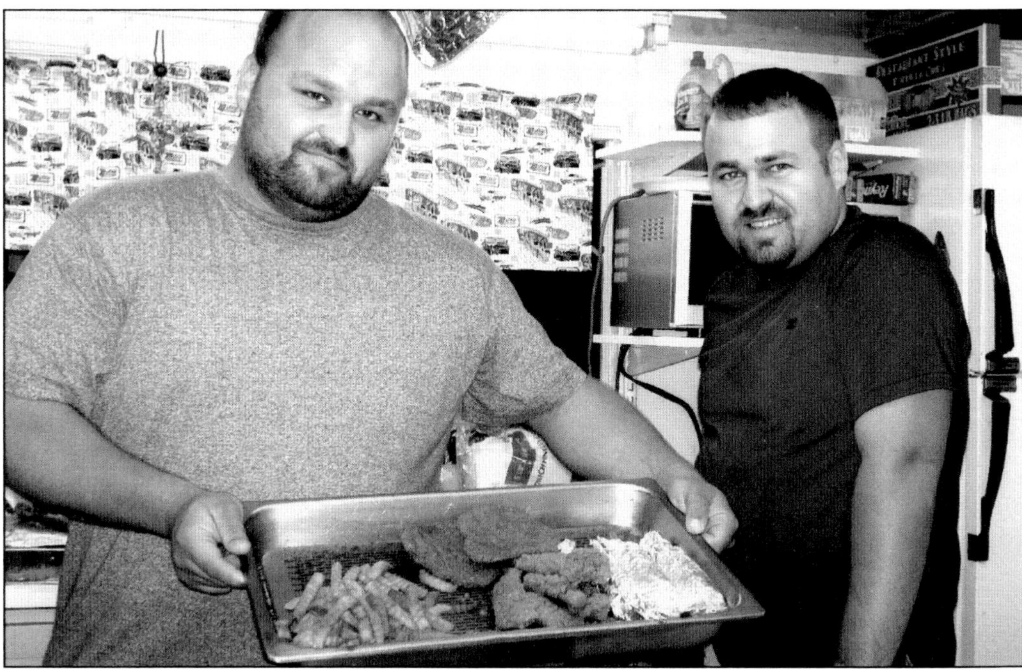

At their food trailer, Jason Craig (left) and Brian Craig (right), of Aurora, offered the crowd next to the main stage what has come to be known as "fair food." (Aurora Lions Club collection.)

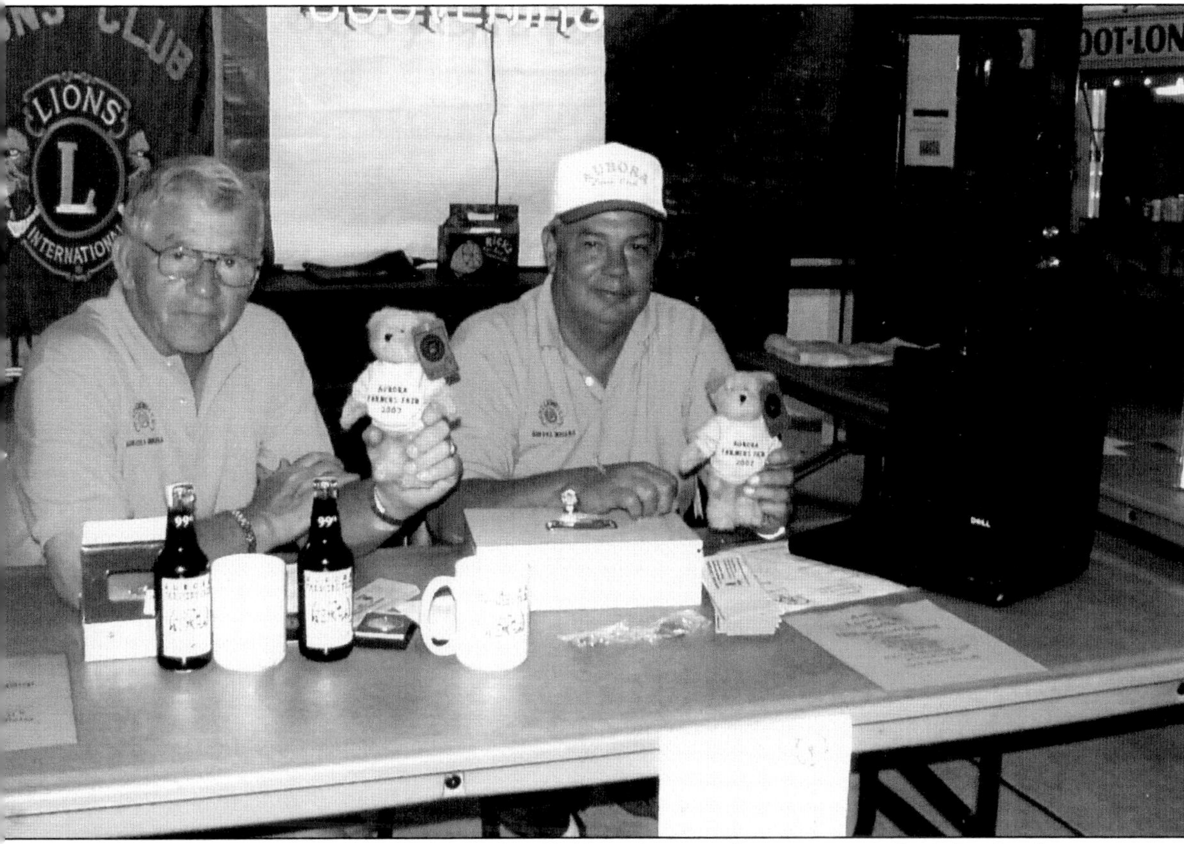

Aurora Lions Club members Howard Aylor (left) and Kenneth Strasemeier (right) take their turn working the souvenir booth. Coffee mugs, 2007 Boyd teddy bears, pop bottles, and pocket knives were some of the imprinted souvenirs. The Aurora Lions Club has promised to make the 100th farmers fair celebration an unforgettable experience. All living queen and king contestants from over the last 100 years will be honored. Many of the old floats and displays will be recreated by area businesses. (Aurora Lions Club collection.)

ACROSS AMERICA, PEOPLE ARE DISCOVERING SOMETHING WONDERFUL. *THEIR HERITAGE.*

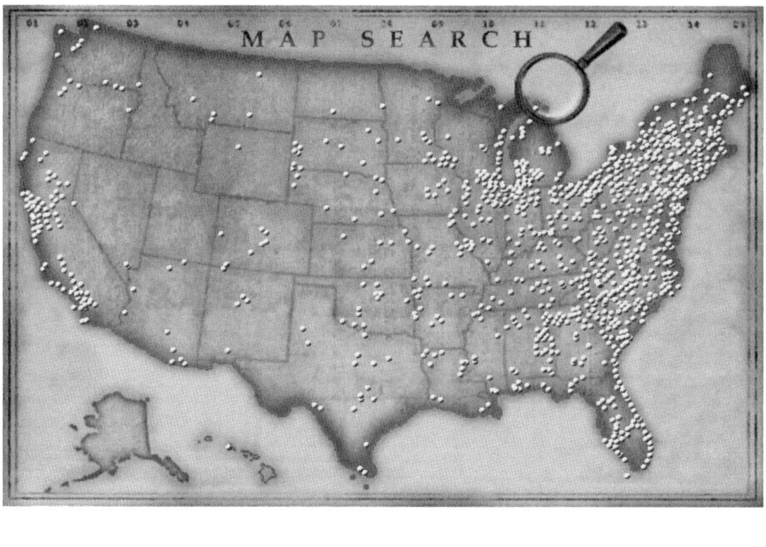